Low-Dimensional Quantum Materials

Low-Dimensional Quantum Materials: From Fundamentals to Frontiers provides a comprehensive exploration of the theoretical foundations, material systems, and emerging applications of quantum materials with reduced dimensionality. From two-dimensional (2D) materials like graphene and transition metal dichalcogenides (TMDs) to Dirac materials, topological insulators, and artificial quantum structures, this book delves into the quantum effects, covering the aspects of electronic, optical, and magnetic properties that arise in these materials.

- Covers fundamental principles, classification, and experimental techniques
- Focuses on specific material systems and their quantum properties
- Highlights frontier applications in quantum computing, optoelectronics, energy, and biomedical technologies
- Addresses theoretical insights and practical challenges such as scalability, integration with existing semiconductor technology, and real-world applicability
- Bridges the gap between fundamental research and technological innovation, providing a roadmap for future advancements in low-dimensional quantum materials

By systematically presenting fundamental concepts, fabrication methods, characterization techniques, and cutting-edge research about low-dimensional quantum materials, this book serves as both an introductory guide and an advanced reference for researchers and professionals in condensed matter physics, nanotechnology, and materials science.

Zongyu Huang is a Professor of the Department of Physics, School of Physics and Optoelectronics, Xiangtan University, China. She earned her B.S. degree in Physics from Hubei University, China, in 2004; her M.S. degree in Condensed Matter Physics from Wuhan University, China, in 2006; and her Ph.D. in Physics from Xiangtan University, China, in 2014. Her research interest is in physical properties and applications of two-dimensional (2D) materials via first principles calculation methods and advanced synthesis and characterization methods. Her work is documented in over 50 peer-reviewed publications in international journals, such as *Advanced Energy Materials* and *Nano Today*.

Long Ren is a Professor at Shanghai University (since December 2024). He previously served as a researcher at Wuhan University of Technology (2021–2024) and as an associate research fellow at the University of Wollongong, Australia (2019–2020). He received his Ph.D. from the University of Wollongong in 2019, his M.S. from Xiangtan University in 2012, and his B.S. from China University of Geosciences (Wuhan) in 2009. His research focuses on interfacial structures and properties of liquid metals and low-dimensional materials, the controlled construction and interface engineering of heterogeneous nanostructures with energy coupling effects, and the fabrication and applications of flexible functional devices.

Jianxin Zhong is a Distinguished Professor of Physics and Executive Director of the Institute for Quantum Science and Technology at Shanghai University. He received his PhD in 1995 from the University of Cergy-Pontoise (now CY Cergy Paris University), France. His research interests span condensed matter physics, computational physics, quantum dynamics and quantum control, as well as the design and growth of advanced quantum materials. He has published more than 400 peer-reviewed papers, which have received over 21,200 citations, with an H-index of 74.

Emerging Materials and Technologies

Series Editor: Boris I. Kharissov

The *Emerging Materials and Technologies* series is devoted to highlighting publications centered on emerging advanced materials and novel technologies. Attention is paid to those newly discovered or applied materials with potential to solve pressing societal problems and improve quality of life, corresponding to environmental protection, medicine, communications, energy, transportation, advanced manufacturing, and related areas.

The series takes into account that, under present strong demands for energy, material, and cost savings, as well as heavy contamination problems and worldwide pandemic conditions, the area of emerging materials and related scalable technologies is a highly interdisciplinary field, with the need for researchers, professionals, and academics across the spectrum of engineering and technological disciplines. The main objective of this book series is to attract more attention to these materials and technologies and invite conversation among the international R&D community.

The Science and Applications of 3D Bioprinting
Edited by Thiago Domingues Stocco and Anderson de Oliveira Lobo

Metal-Organic Frameworks for Anticorrosive Coatings
Edited by Ashish Kumar, Ambrish Singh, and Humira Assad

Evaluation of Electrochemical Corrosion: Experimental Procedures and Characterization
Nam Nguyen-Dang

Smart Functional Wearables: From Nanofibers to Textiles
Majid Montazer and Ali Mianehro

Waste Valorization to Nanomaterials: Classification, Synthesis Methods, and Applications
Anele Mpupa, Shirley Kholofelo Selahle, Philiswa Nosizo Nomngongo, Mthokozisi Mnguni, Ngwako Joseas Waleng, and Siyabonga Shoba

Handbook of Two-Dimensional Quantum Materials and Energy Applications
Edited by Abdus Salam Sarkar and Emmanuel Stratakis

Low-Dimensional Quantum Materials: From Fundamentals to Frontiers
Edited by Zongyu Huang, Long Ren, and Jianxin Zhong

For more information about this series, please visit:
www.routledge.com/Emerging-Materials-and-Technologies/book-series/CRCEMT

Low-Dimensional Quantum Materials

From Fundamentals to Frontiers

Edited by
Zongyu Huang
Long Ren, and
Jianxin Zhong

CRC Press
Taylor & Francis Group
Boca Raton London New York

CRC Press is an imprint of the
Taylor & Francis Group, an **informa** business

Designed cover image: © 2026 Shutterstock

First edition published 2027
by CRC Press
2385 NW Executive Center Drive, Suite 320, Boca Raton FL 33431

and by CRC Press
4 Park Square, Milton Park, Abingdon, Oxon, OX14 4RN

CRC Press is an imprint of Taylor & Francis Group, LLC

ISBN: 978-1-041-10485-8 (hbk)
ISBN: 978-1-041-10486-5 (pbk)
ISBN: 978-1-003-65530-5 (ebk)

DOI: 10.1201/9781003655305

Typeset in Times
by SPi Technologies India Pvt Ltd (Straive)

Contents

Part III Cutting-Edge Research and Future Applications

Preface

The rapid advancements in quantum materials over the past few decades have transformed our understanding of condensed matter physics and nanotechnology. Among these, low-dimensional quantum materials have emerged as a distinct class of systems with extraordinary electronic, optical, and mechanical properties. These materials, ranging from atomically thin 2D materials such as graphene and TMDs to low-dimensional topological insulators, quantum wires, and artificial quantum structures, hold immense promise for next-generation electronic, optoelectronic, spintronic, and quantum computing applications.

This book, *Low-Dimensional Quantum Materials: From Fundamentals to Frontiers*, aims to provide a comprehensive and systematic introduction to the field. It is structured to serve as both an entry point for newcomers and a reference for researchers and professionals actively working in material science, physics, and engineering. The book is divided into three main parts:

"Part I: Introduction to Low-Dimensional Quantum Materials" – This section lays the theoretical foundation by defining low-dimensional quantum materials and highlighting their unique properties compared to traditional three-dimensional systems. It covers the classification, synthesis, and characterization techniques used to explore these novel materials. Fundamental quantum phenomena such as quantum confinement, topological effects, and quantum phase transitions are also discussed.

"Part II: Major Systems of Low-Dimensional Quantum Materials" – The second section focuses on key material families, including graphene and its derivatives, transition metal chalcogenides, low-dimensional topological and superconducting materials, magnetic materials, and artificially engineered quantum systems. The discussion extends to their structural properties, electronic behaviors, and potential functionalities in various technologies.

"Part III: Cutting-Edge Research and Future Applications" – The final section explores the role of low-dimensional quantum materials in emerging technologies such as quantum computing, neuromorphic devices, energy storage and catalysis, environmental monitoring, and biomedicine. It also highlights key challenges in material synthesis, scalability, integration with existing semiconductor technology, and commercialization.

Contributors

Xinchun Du
Institute for Quantum Science and Technology
Shanghai University
Shanghai, China

Biwen He
School of Physics and Optoelectronics
Xiangtan University
Xiangtan, China

Xixi Huang
Institute for Quantum Science and Technology
Shanghai University
Shanghai, China

Zongyu Huang
School of Physics and Optoelectronics
Xiangtan University
Xiangtan, China

Jie Li
Institute for Quantum Science and Technology
Shanghai University
Shanghai, China

Junjie Liu
Institute for Quantum Science and Technology
Shanghai University
Shanghai, China

Yundan Liu
School of Physics and Optoelectronics
Xiangtan University
Xiangtan, China

Siwei Luo
School of Physics and Optoelectronics
Xiangtan University
Xiangtan, China

Dan Mu
Institute for Quantum Science and Technology
Shanghai University
Shanghai, China

Xiang Qi
School of Physics and Optoelectronics
Xiangtan University
Xiangtan, China

Long Ren
Institute for Quantum Science and Technology
Shanghai University
Shanghai, China

Lamei Tan
School of Physics and Optoelectronics
Xiangtan University
Xiangtan, China

Wei Chen
School of Physics and Optoelectronics
Xiangtan University
Xiangtan, China

Yi Zhang
Institute for Quantum Science and Technology
Shanghai University
Shanghai, China

Jianxin Zhong
Institute for Quantum Science and Technology
Shanghai University
Shanghai, China

Haoyan Zhou
School of Physics and Optoelectronics
Xiangtan University
Xiangtan, China

Lihong Zhou
Institute for Quantum Science and Technology
Shanghai University
Shanghai, China

Part I

Introduction to Low-Dimensional Quantum Materials

1 Overview of Low-Dimensional Quantum Materials

Xixi Huang, Long Ren, Zongyu Huang, and Jianxin Zhong

1.1 INTRODUCTION

As Richard Feynman famously remarked in 1982,[1] "Nature isn't classical, and if you want to make a simulation of nature, you'd better make it quantum mechanical." This insight underpins quantum information science, which seeks to exploit uniquely quantum phenomena to process, transmit, and store information beyond the capabilities of classical systems.[2–4] In recent years, significant progress has been made toward realizing quantum computers. Notably, Google AI Quantum demonstrated a programmable superconducting processor ("Sycamore") that performed a task beyond the practical reach of classical supercomputers, marking a milestone often referred to as "quantum supremacy,"[5] as illustrated in Figure 1.1. At the heart of this endeavor are quantum bits (qubits), which harness superposition, entanglement, and coherence to encode and manipulate information in ways that classical bits fundamentally cannot process.[6–8] However, the realization of such quantum devices hinges on materials that can sustain and control coherent quantum states, which positions quantum materials at the forefront of the emerging quantum technology landscape.[9,10]

The realization of quantum technologies relies on quantum materials capable of hosting and coherently manipulating quantum states with high fidelity and long coherence times.[11–13] Among the diverse material platforms explored, low-dimensional quantum materials have attracted particular attention due to their pronounced quantum confinement effects.[14–17] Early studies on low-dimensional quantum materials primarily focused on theoretical calculations and fundamental property exploration, where models and first-principles calculations predict attractive phenomena such as topological states, two-dimensional (2D) magnetism, and low-dimensional superconductivity. For example, Eugene J. Mele introduced the Z_2 topological invariant to characterize the quantum spin Hall phase in time-reversal-invariant systems, laying the theoretical foundation for 2D topological insulators.[18] With the rapid progress in materials synthesis techniques, including mechanical exfoliation, molecular beam epitaxy (MBE), and chemical vapor deposition (CVD), the controlled fabrication of high-quality 2D crystals and van der Waals heterostructures has become possible.

DOI: 10.1201/9781003655305-2

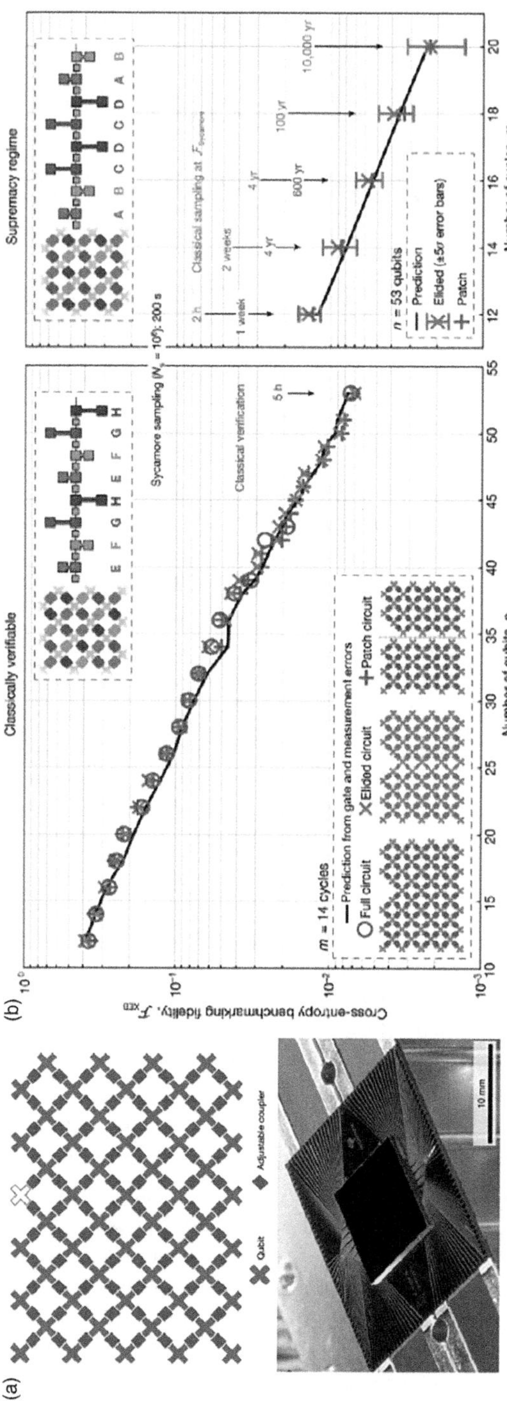

FIGURE 1.1 Quantum computer. (a) The Sycamore processor. (b) Demonstrating quantum supremacy.[5]

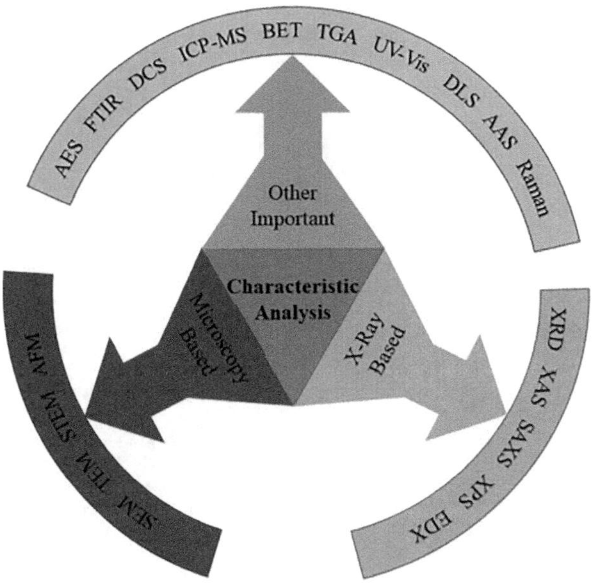

FIGURE 1.2 Characterization tools for nanomaterial characterization.[25]

Such achievements thereby establish robust platforms for experimental investigations of low-dimensional quantum phenomena.[19–24]

Meanwhile, advances in characterization techniques also played a crucial role in revealing the structural and electronic properties of low-dimensional quantum materials (Figure 1.2). Transmission electron microscopy, first developed by Ernst Ruska in the 1930s, enables atomic-scale imaging of crystal structures and has been widely used to resolve lattice configurations, defects, and stacking orders in 2D materials and van der Waals heterostructures.[26] The invention of scanning tunneling microscopy (STM) by Gerd Binnig and Heinrich Rohrer enabled direct real-space observation of surfaces with atomic resolution. It enabled probing of the local electronic density of states via tunneling spectroscopy.[27,28] Combined with complementary techniques such as angle-resolved photoemission spectroscopy, Raman spectroscopy, and magneto-transport measurements, precise determination of electronic band structures, magnetic ordering, and topological characteristics in low-dimensional systems could be realized.[29–31] These characterization tools not only reveal structural and electronic details at the atomic scale but also provide crucial insight into how reduced dimensionality alters the fundamental properties of materials.

In low-dimensional materials, spatial confinement imposes boundary conditions on the electron wavefunctions, leading to the quantization of energy levels and the emergence of discrete or reduced-dimensional electronic states (Figure 1.3). Reduced dimensionality enhances quantum confinement and interaction effects, often giving rise to quantum phenomena that are absent in bulk systems. When charge carriers are confined to length scales comparable to their de Broglie wavelengths, the resulting quantum confinement reshapes the band structure and density of states. The enhanced

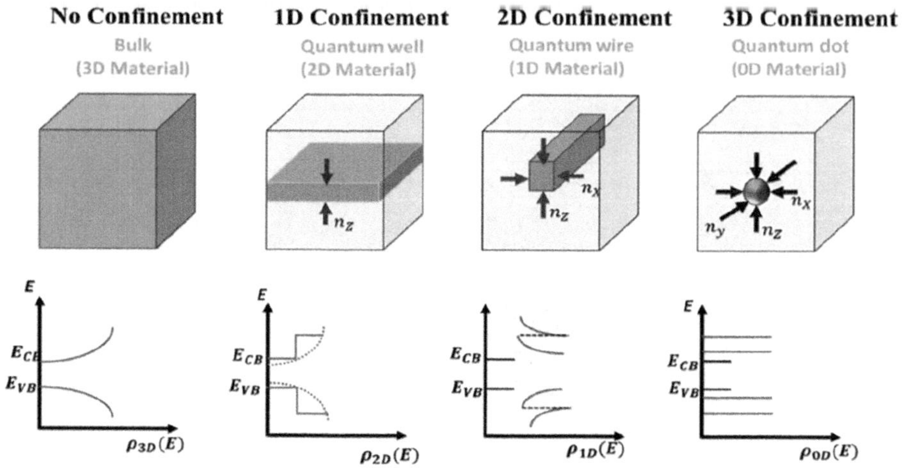

FIGURE 1.3 Quantum motion constraints across different dimensions.[13]

Coulomb interactions and reduced dielectric screening could further amplify many-body effects.[32–35]

These fundamental effects manifest in a variety of representative low-dimensional material systems, giving rise to diverse emergent quantum phenomena. For example, as demonstrated in the pioneering work by A. K. Geim and K. S. Novoselov, the linear band dispersion near the K and K′ points in graphene leads to the emergence of massless Dirac fermions, which give rise to unusual transport properties such as extremely high carrier mobility and the anomalous quantum Hall effect.[31] Unconventional superconductivity emerges in magic-angle twisted bilayer graphene, where ultra-flat moiré bands generated by a ~1.1° interlayer twist produce strong electronic correlations.[36] In 2D semiconductors such as molybdenum diselenide ($MoSe_2$) and tungsten diselenide (WSe_2), reduced dielectric screening and strong quantum confinement lead to exceptionally large exciton binding energies reaching hundreds of millielectronvolts. It would enable the observation of robust excitonic and valley-dependent optical phenomena.[37,38] Meanwhile, in low-dimensional topological systems, strong spin–orbit coupling and band inversion can give rise to topologically protected edge states with spin–momentum locking. For example, the quantum spin Hall effect was experimentally observed in mercury telluride/cadmium telluride (HgTe/CdTe) quantum wells by Laurens W. Molenkamp et al., providing the first realization of a 2D topological insulator.[39] Collectively, these findings illustrate that quantum confinement and many-body interactions often give rise to novel states that are absent in bulk counterparts, forming the foundation for both fundamental studies and future device applications.

In this book, we present a comprehensive and systematic treatment of low-dimensional quantum materials, aiming to bridge foundational concepts and the current frontiers of research and applications. The subsequent chapters will introduce classification schemes, synthesis and characterization methods, theoretical frameworks, and representative physical phenomena, providing a resource that serves both newcomers entering the field and active researchers seeking a coherent overview of this rapidly evolving discipline.

1.2 CLASSIFICATION OF LOW-DIMENSIONAL QUANTUM MATERIALS

Low-dimensional quantum materials encompass a wide spectrum of systems with diverse structural motifs and emergent quantum phenomena. To systematically capture their fundamental properties and technological potential, several complementary classification schemes have been developed, typically based on dimensionality and intrinsic physical characteristics. These frameworks not only provide distinct perspectives for understanding the underlying physics but also offer practical guidance for identifying and engineering materials tailored to next-generation quantum technologies, including quantum computing, spintronics, and advanced optoelectronic devices.

Low-dimensional materials are typically divided into zero-dimensional (0D), one-dimensional (1D), and 2D systems,[40,41] as shown in Figure 1.4. In 0D materials such as quantum dots and nanocrystals, carriers are confined in all three spatial directions, resulting in discrete energy levels similar to those of atoms. These materials often exhibit strong quantum confinement effects and size-dependent optical properties. A recent example was reported by Maksym V. Kovalenko et al., who demonstrated single-photon superradiance in perovskite quantum dots, revealing ultrafast radiative decay and coherent light emission that highlights their potential as quantum light sources for quantum communication and photonic quantum technologies.[42] Notably, 1D materials, including nanowires, nanotubes, and quantum wires, are confined in two spatial directions but allow carrier motion along a single axis. Such reduced dimensionality often gives rise to pronounced quantum confinement and coherent transport phenomena. For example, Leo P. Kouwenhoven et al. performed tunneling

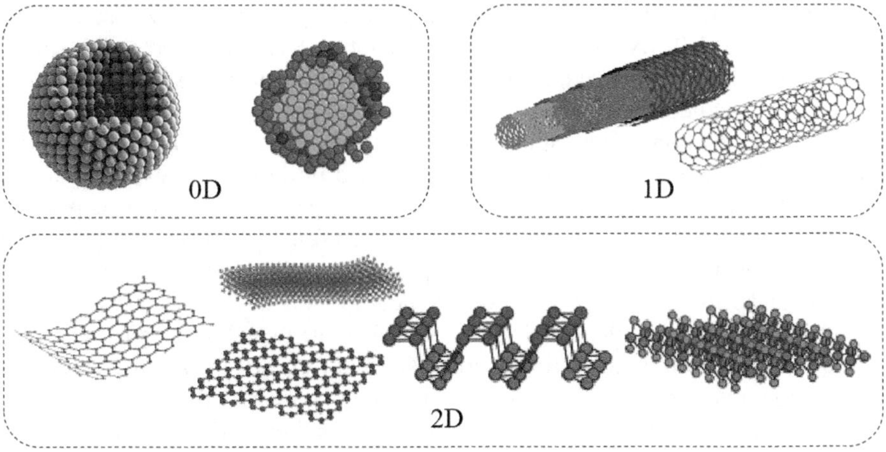

FIGURE 1.4 Classification of low-dimensional quantum materials by dimensions.

spectroscopy measurements on hybrid devices composed of an indium antimonide nanowire coupled to a superconductor. Under an external magnetic field, a robust zero-bias conductance peak was observed, providing experimental evidence for Majorana zero modes and establishing an important platform for 1D topological quantum computing.[43] Notably, 2D materials represent another important category of low-dimensional systems, in which carriers are free to move within an atomic plane but are confined in the perpendicular direction. Typical examples include graphene, transition–metal dichalcogenides (TMDs), and other layered van der Waals materials. Due to their atomic thickness and reduced dimensionality, these materials often exhibit high carrier mobility, strong light–matter interactions, and tunable electronic structures. For instance, Qiaoling Lin et al. recently showed in molybdenum disulfide (MoS_2)/WSe_2 hetero-bilayers that moiré-engineered potentials can localize interlayer excitons and significantly modify light–matter interactions at room temperature, demonstrating a highly tunable optical response in atomically thin TMD systems.[44]

Low-dimensional quantum materials can be systematically classified by their emergent physical properties, reflecting different quantum phases that arise when electronic, spin, and topological degrees of freedom are constrained by reduced dimensionality. One central class is low-dimensional topological materials, where nontrivial band topology gives rise to robust boundary states that are stable against perturbations preserving the relevant symmetry. In low-dimensional systems, 2D magnetic topological insulators exhibit exotic phases, such as antiferromagnetic topological insulators, in which nontrivial band topology coexists with intrinsic magnetism, leading to quantized edge transport even in the absence of an external magnetic field. Monolayer CrI_3, a 2D van der Waals magnet, has been demonstrated to realize a second-order topological phase featuring robust corner states, highlighting the interplay between intrinsic magnetism and nontrivial band topology.[45] In addition, low-dimensional superconducting materials exhibit characteristic signatures such as zero resistance and the Meissner effect. In certain systems, superconductivity can coexist with a nontrivial band structure, realizing topological superconductivity that supports Majorana edge modes and provides a potential platform for fault-tolerant quantum computation. A notable experimental example is the van der Waals heterostructure $CrBr_3$/$NbSe_2$, where low-temperature STM and spectroscopy reveal signatures of 1D Majorana edge modes in a 2D superconducting system, as demonstrated by S. Kezilebieke.[46] Finally, low-dimensional van der Waals magnetic materials exhibit robust long-range spin order and strong magnetic anisotropy, enabling emergent phenomena such as ferromagnetism and intricate spin textures that intimately influence electronic transport and topological properties. A recent experimental demonstration by Deyi Fu shows that Co-doped few-layer black phosphorus can sustain ferromagnetism up to room temperature while maintaining high carrier mobility and semiconducting behavior. By incorporating this material into magnetic tunnel junction devices, the study further reveals large tunneling magnetoresistance at room temperature, highlighting a promising platform for integrating magnetism with charge transport in low-dimensional spintronic devices.[47] These property-based categories, exemplified by topological, superconducting, and magnetic materials, offer a conceptual lens for understanding the complex quantum phases in low-dimensional systems.

Overall, the classification of low-dimensional quantum materials from multiple perspectives provides a comprehensive framework for understanding their diverse properties and behaviors. While dimensionality offers a fundamental description of the spatial confinement of carriers, classification based on physical properties is particularly valuable for identifying the functional characteristics of these materials and guiding their potential applications. As research in this field continues to advance, new classes of low-dimensional quantum materials and hybrid systems are likely to emerge, further enriching the landscape of quantum materials science.

1.3 RESEARCH DEVELOPMENT OF LOW-DIMENSIONAL QUANTUM MATERIALS

The exploration of low-dimensional quantum materials can be traced back to early studies of semiconductor quantum structures in the late twentieth century. In particular, the discovery of the quantum Hall effect by Klaus von Klitzing in 2D electron gases formed in semiconductor heterostructures provided one of the first clear demonstrations of quantum phenomena arising from reduced dimensionality.[48] Subsequent advances in nanotechnology enabled the fabrication and investigation of other low-dimensional systems, including 1D carbon nanotubes and 0D quantum dots, which further revealed rich quantum confinement effects.[49] A major breakthrough occurred with the isolation of graphene, demonstrating that stable atomically thin crystals could exist and exhibit extraordinary electronic and mechanical properties.[50] This discovery triggered a surge of interest in 2D materials and stimulated extensive research on a wide variety of layered systems, including TMDs and van der Waals heterostructures.[51–53] Together, these developments have established low-dimensional quantum materials as a central platform for exploring novel quantum phenomena and developing next-generation quantum technologies.

As research on low-dimensional quantum materials has advanced, increasing effort has been devoted to actively tuning their quantum properties through external stimuli. Electrostatic gating represents one of the most direct and powerful approaches. In pioneering experiments by Konstantin S. Novoselov and Andre K. Geim, the application of a gate voltage to graphene devices enabled continuous control of carrier density and revealed unconventional quantum Hall states in a 2D Dirac system.[50] Similar electrical control has also been demonstrated in atomically thin semiconductors. Field-effect transistors based on monolayer MoS_2, developed by Andras Kis et al., demonstrated that electrostatic gating can efficiently modulate carrier transport and switch the material between insulating and conductive regimes.[56] Mechanical deformation provides another effective tuning parameter. Experiments have shown that tensile strain applied to monolayer TMDs such as MoS_2 can significantly modify their band structures and optical transitions, highlighting the strong coupling between lattice distortion and electronic states in atomically thin crystals (Figure 1.5).[57,58]

In addition, magnetic and optical fields provide powerful means to manipulate spin, valley, and excitonic degrees of freedom in 2D systems. Magnetic fields, for

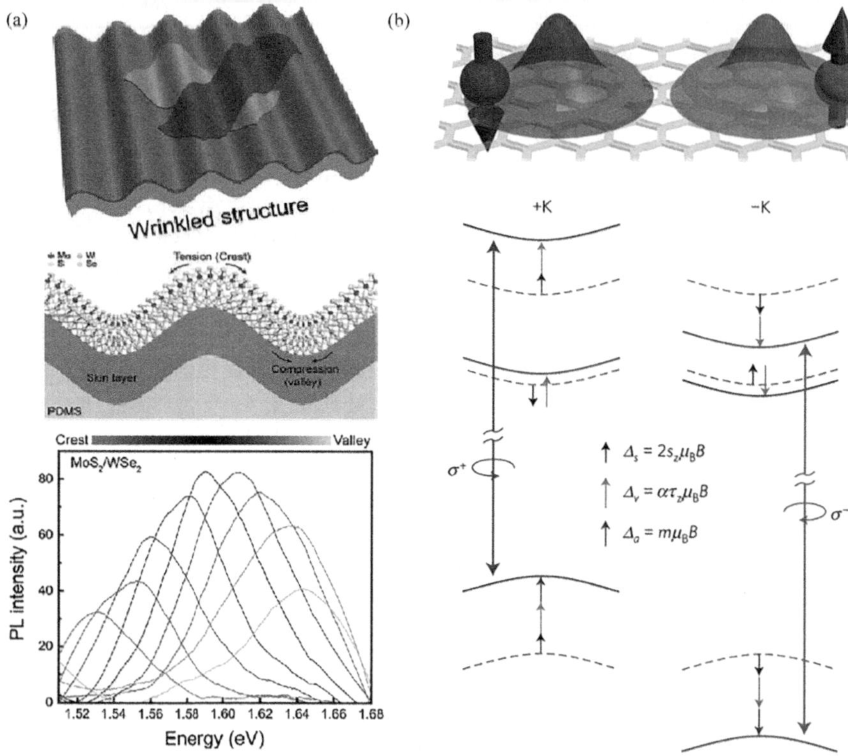

FIGURE 1.5 Field-controlled low-dimensional quantum materials. (a) Strain engineering of MoS_2/WSe_2 hetero-bilayers.[54] (b) Magnetic control of valley pseudospin in monolayer WSe_2.[55]

instance, can lift the degeneracy of valley states in TMDs. In monolayer WSe_2, Xu et al. demonstrated that an external magnetic field induces a large valley Zeeman splitting, providing direct control over valley pseudospin and enabling investigations of spin–valley coupling.[55] Optical fields offer another complementary approach. Using circularly polarized light, Kin Fai Mak et al. selectively excited carriers in specific valleys of monolayer MoS_2, enabling manipulation of valley polarization.[59] Together, these studies illustrate that external magnetic and optical fields can be exploited to systematically manipulate quantum degrees of freedom in atomically thin materials, providing essential tools for both fundamental studies and potential valleytronic or spintronic applications.

On the other hand, the weak interlayer Van der Waals coupling in layered materials provides exceptional flexibility for assembling artificial heterostructures, as shown in Figure 1.6. For example, Cory Dean et al. showed that graphene aligned on hexagonal boron nitride (hBN) exhibits moiré patterns that substantially modify the electronic spectrum and induce secondary Dirac points.[61] Twisted bilayer graphene near the so-called “magic angle” was demonstrated by Cao to host correlated insulating states and unconventional superconductivity, highlighting the role of interlayer coupling and electron correlations in low-dimensional systems.[36] Similar

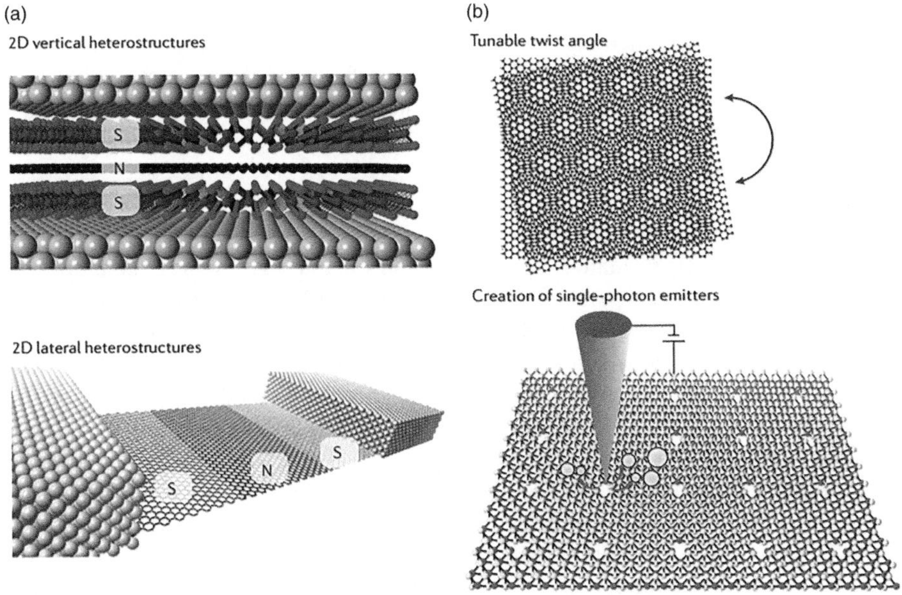

FIGURE 1.6 Artificial preparation of low-dimensional quantum materials for manipulating quantum properties.[60] (a) Bottom-up heterostructures. (b) Controlled fabrication.

effects have been observed in TMD heterostructures such as $MoSe_2/WSe_2$, where Xiaodong Xu et al. reported the formation of interlayer excitons with tunable valley polarization arising from interlayer hybridization.[62] These strategies collectively enable precise tuning of low-dimensional quantum materials, providing controllable platforms to realize and harness quantum phenomena for next-generation quantum technologies.

With the continuous advancement of technology, low-dimensional quantum materials have gradually transitioned from fundamental research toward a range of emerging technological applications. For instance, superconducting circuits based on low-dimensional superconducting materials have become one of the leading platforms for quantum information processing. William D. Oliver et al. integrated van der Waals Josephson junctions based on 2D materials into a transmon circuit architecture to realize hBN-based transmon qubits, demonstrating high-coherence superconducting quantum circuits.[63] Another representative example is the work by Tongcang Li et al., who realized optical polarization and coherent control of nearby nitrogen nuclear spins using boron-vacancy defects in hBN.[64] Electron-spin-mediated coupling further enabled strong interactions between nuclear spins, providing an important experimental foundation for scalable nuclear-spin qubit platforms in 2D van der Waals materials. Meanwhile, the reduced dimensionality and strong spin–orbit coupling present in many 2D materials also make them attractive for spintronic and optoelectronic applications. A representative example is the work by K. F. Mak et al., who demonstrated the valley Hall effect in monolayer MoS_2 transistors.[65] By illuminating the devices with circularly polarized light, they selectively excited electrons into specific valleys, producing an anomalous Hall voltage with a sign determined by the light helicity, thereby establishing the valley degree of freedom as a controllable information carrier in 2D materials. Together, these examples illustrate how low-dimensional quantum materials harness their unique electronic, spin, and valley degrees of freedom to enable advanced quantum information, spintronic, and optoelectronic functionalities.

In conventional electronics, atomically thin semiconductors have been used to develop high-performance photodetectors, light-emitting devices, and ultrathin solar cells. Their strong light–matter interactions and tunable band gaps enable efficient optical absorption and emission in nanoscale devices. Wang demonstrates that monocrystalline MoS_2/polycrystalline rhenium disulfide (ReS_2) lateral heterojunctions enable high-performance photodetectors with exceptional responsivity and quantum efficiency.[66] Low-dimensional materials are also promising for flexible and wearable electronics, as their atomic-scale thickness and mechanical flexibility enable integration into bendable, transparent electronic systems. Chen et al. demonstrate a monolayer graphene-based flexible photodetector with asymmetric metal electrodes.[67] The device exploits the photo-thermoelectric effect to achieve ultra-broadband operation from 325 nm (UV) to 1.19 mm (MM), with a responsivity of ~396 mV W^{-1}, and a fast ~0.8 ms response time, highlighting its promise for wearable optoelectronics.

Overall, the research development of low-dimensional quantum materials has evolved from the discovery of individual materials such as graphene to the exploration of a vast family of quantum systems with diverse properties and tunable functionalities. The combination of novel material platforms, controllable physical properties, and rapidly expanding technological applications continues to drive the field forward. As experimental techniques and theoretical models continue to improve, low-dimensional quantum materials are expected to play an increasingly important role in both fundamental physics and next-generation electronic and quantum technologies.

1.4 LOW-DIMENSIONAL QUANTUM MATERIALS IN THE FUTURE

Looking forward, the future development of low-dimensional quantum materials is expected to increasingly rely on the interdisciplinary integration of materials science, quantum information, and artificial intelligence. The discovery and optimization of quantum materials have traditionally relied on empirical exploration and theoretical prediction. However, the rapid growth of computational capabilities and data-driven approaches is transforming this paradigm toward more efficient and predictive materials design.

Artificial intelligence and machine learning techniques provide powerful tools for accelerating the discovery of novel low-dimensional materials and predicting their electronic, magnetic, and topological properties. A representative example is the work by Jing Wang et al., who employed supervised neural networks combined with first-principles calculations to screen 2D material databases for magnetic topological insulators.[68] Their model identified six new classes comprising 15 candidate Chern insulators, among which four classes (seven materials) possess full topological band gaps, providing promising targets for experimental realization of quantum anomalous Hall states. Liang Fu et al., who introduced a machine-learned descriptor termed "topogivity" to evaluate the tendency of elements to form topological materials and employed this rule to screen large material databases, leading to the discovery of several new topological material candidates beyond those identifiable by conventional symmetry indicators.[69] By combining high-throughput first-principles calculations with data-driven models, researchers can rapidly screen large material databases and identify candidate systems with desirable quantum characteristics. Such approaches enable the rational design of materials with tailored band structures, strong correlations, or nontrivial topological phases, significantly shortening the time required for experimental realization.

In parallel, the development of quantum information science is generating ever-growing demands for materials with precisely controllable quantum degrees of freedom. Low-dimensional platforms, including 2D superconductors, van der Waals heterostructures, and defect-engineered quantum systems, offer versatile opportunities for implementing quantum bits, sensors, and communication devices. Figure 1.7 shows the detection and coherent manipulation of a single nuclear spin using boron-vacancy defects in hBN.[70] These optically addressable defect spins enable atomic-scale NMR and quantum sensing, establishing defect centers in 2D van der Waals materials as promising solid-state qubit platforms. Similarly, Liu et al. reported spin defects in the 2D material β-GeS_2, achieving coherent control with spin coherence times on the order of tens of microseconds.[71] The substantially enhanced coherence relative to hBN defects underscores β-GeS_2 as a compelling host for long-coherence defect-based qubits. In a complementary approach, Switzer et al. employed STM to precisely position two titanium atoms on a MgO/Ag(100) surface, constructing a two-qubit system and generating entangled spin states (Bell states).[72] This demonstration exemplifies the potential of STM-based atomic engineering to realize quantum materials tailored for qubit implementation. Taken together, these examples highlight the synergy between predictive materials modeling, precise fabrication techniques, and quantum characterization, pointing toward the emerging concept of "materials by design," in which targeted quantum functionalities can be realized through predictive modeling, automated synthesis, and closed-loop optimization strategies. By integrating advanced characterization tools with intelligent data analysis, future research may establish autonomous platforms capable of continuously optimizing material synthesis and device performance for quantum information applications.

Conversely, low-dimensional quantum materials, owing to their atomic-scale thickness, mechanical flexibility, and tunable quantum properties, hold significant promise for next-generation flexible quantum devices and bio-integrated quantum technologies. Their mechanical compliance enables incorporation into bendable or wearable quantum sensors, capable of detecting magnetic fields, temperature, or biochemical signals at the nanoscale.[67,73,74] Moreover, low-dimensional materials have rapidly advanced beyond fundamental physics into bioconverged applications, particularly in biosensing and wearable biomedical devices. Graphene, TMDs, MXenes, and other atomically thin materials exhibit high surface area, excellent electrical conductivity, mechanical flexibility, and tunable surface chemistry, making them ideal candidates for real-time biological analyte detection and body-integrated sensors.[75–77]

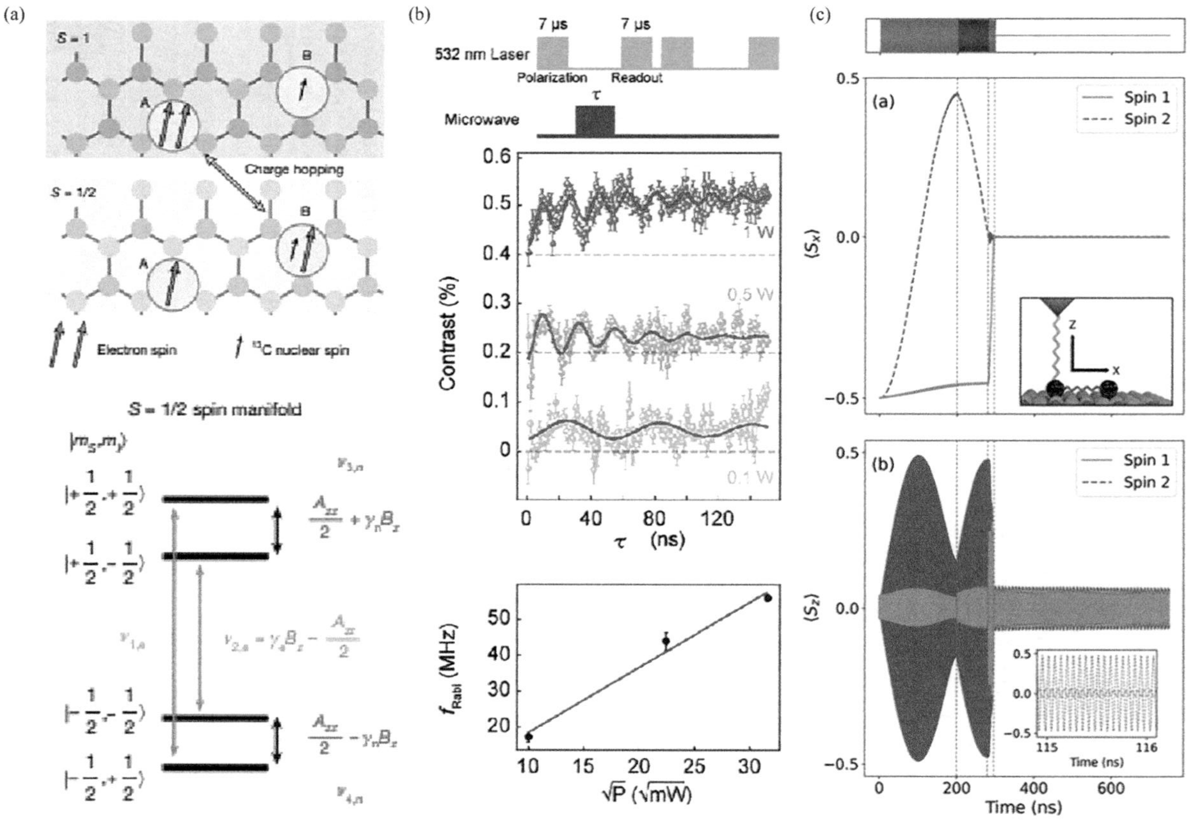

FIGURE 1.7 Qubit preparation using low-dimensional materials. (a) Observation of three types of single-spin defects in hBN.[70] (b) Coherent manipulation of spin defects in β-GeS_2.[71] (c) Spin evolution during the quantum circuit execution.[72]

By combining highly coherent spin defects with flexible substrates or biocompatible interfaces, these materials could support implantable quantum sensors or bioelectronic devices that operate under physiological conditions. Such integration of low-dimensional quantum materials with soft and biological systems represents a transformative frontier, bridging quantum information science, flexible electronics, and bioengineering. It could enable applications ranging from neural signal detection to real-time molecular diagnostics.

The intersection of materials science, quantum information, and artificial intelligence is poised to accelerate the identification of novel low-dimensional quantum materials and to facilitate their seamless translation from fundamental research to functional quantum technologies. As both fabrication methodologies and theoretical frameworks advance, low-dimensional quantum systems are expected to assume a central role in driving the evolution of quantum science and enabling next-generation electronic and quantum devices.

1.5 SUMMARY

In this chapter, we have presented a comprehensive overview of low-dimensional quantum materials, covering their fundamental concepts, classification schemes, historical evolution, and emerging research directions. These materials are characterized by confinement of at least one spatial dimension to the nanoscale, resulting in reduced dimensionality and pronounced quantum confinement effects. Such dimensional reduction fundamentally alters the electronic band structure, density of states, and optical and transport properties compared with conventional three-dimensional systems.[32,33] As a result, a wide range of novel quantum phenomena can emerge in these systems, including Dirac fermions, strongly bound excitons, topologically protected states, and gate-tunable superconductivity, establishing low-dimensional materials as versatile platforms for exploring emergent quantum physics.[31,36,39]

We discussed systematic classification schemes that provide a clear framework for understanding low-dimensional quantum materials from both structural and functional perspectives. Dimensionality-based classification categorizes these materials into 0D, 1D, and 2D systems, reflecting the degree of spatial confinement and its impact on carrier behavior and density of states. Complementarily, classification based on physical properties, including electronic, optical, magnetic, and topological characteristics, captures the diverse quantum functionalities exhibited by these systems. This dual perspective not only clarifies the fundamental distinctions among different classes of low-dimensional materials but also highlights their broad potential for applications in electronics, photonics, spintronics, and quantum information technologies.

Low-dimensional quantum materials are expected to play an increasingly pivotal role in the advancement of quantum science and technology. Defect engineering and atomic-scale fabrication techniques, such as STM, enable the deterministic creation of tailored quantum states and qubit architectures with atomic precision.[72] In contrast, the integration of predictive theoretical approaches and data-driven methods is accelerating the paradigm of "materials by design," allowing targeted quantum functionalities to be realized with greater efficiency.[68,69] The intrinsic mechanical flexibility and surface sensitivity of low-dimensional systems open new opportunities for their incorporation into flexible, wearable, and bio-integrated quantum devices, expanding their application scope beyond conventional solid-state platforms.[73,76] Together, these developments highlight a clear trajectory from fundamental discovery toward programmable, application-oriented quantum materials, positioning low-dimensional systems at the forefront of next-generation quantum technologies.

Overall, this chapter establishes the conceptual foundation for the remainder of the book. In the subsequent part, we will first focus on the preparation strategies and advanced characterization techniques that enable precise control and reliable probing of low-dimensional systems, followed by a detailed discussion of their fundamental physical properties, including electronic, optical, and magnetic behaviors. Building on these foundations, the second part of the book will introduce the major classes of low-dimensional quantum materials, providing representative systems and their characteristic properties. Finally, the third part will highlight cutting-edge research directions and emerging applications, with particular emphasis on quantum information technologies, novel device architectures, and interdisciplinary integration. Together, these sections form a coherent progression from fundamental understanding to material realization and ultimately to advanced functionalities and practical applications.

REFERENCES

[1] Feynman, R. P., Simulating physics with computers. *International Journal of Theoretical Physics* 21, 467–488 (1982).

[2] Montanaro, A., Quantum algorithms: an overview. *NPJ Quantum Information* 2, 15023 (2016).

[3] Long, G.-L., Progress in quantum information. *National Science Review* 12, nwaf289 (2025).

[4] Head-Marsden, K., Flick, J., Ciccarino, C. J., et al., Quantum information and algorithms for correlated quantum matter. *Chemical Reviews* 121, 3061–3120 (2021).

[5] Arute, F., Arya, K., Babbush, R., et al., Quantum supremacy using a programmable superconducting processor. *Nature* 574, 505–510 (2019).

[6] Siddiqi, I., Engineering high-coherence superconducting qubits. *Nature Reviews Materials* 6, 875–891 (2021).

[7] Clarke, J., Wilhelm, F. K., Superconducting quantum bits. *Nature* 453, 1031–1042 (2008).

[8] Renner, M. J., Tavakoli, A., Quintino, M. T., Classical cost of transmitting a qubit. *Physical Review Letters* 130, 120801 (2023).

[9] Keimer, B., Moore, J. E., The physics of quantum materials. *Nature Physics* 13, 1045–1055 (2017).

[10] Cava, R., de Leon, N., Xie, W., Introduction: Quantum materials. *Chemical Reviews* 121, 2777–2779 (2021).

[11] Awschalom, D. D., Hanson, R., Wrachtrup, J., et al., Quantum technologies with optically interfaced solid-state spins. *Nature Photonics* 12, 516–527 (2018).

[12] Devoret, M. H., Schoelkopf, R. J., Superconducting circuits for quantum information: An outlook. *Science* 339, 1169–1174 (2013).

[13] Goyal, R. K., Maharaj, S., Kumar, P., et al., Exploring quantum materials and applications: A review. *Journal of Materials Science: Materials in Engineering* 20, 4 (2025).

[14] Manzeli, S., Ovchinnikov, D., Pasquier, D., et al., 2D transition metal dichalcogenides. *Nature Reviews Materials* 2, 17033 (2017).
[15] Glavin, N. R., Rao, R., Varshney, V., et al., Emerging applications of elemental 2D materials. *Advanced Materials* 32, 1904302 (2020).
[16] Novoselov, K. S., Mishchenko, A., Carvalho, A., et al., 2D materials and van der Waals heterostructures. *Science* 353, aac9439 (2016).
[17] Butler, S. Z., Hollen, S. M., Cao, L., et al., Progress, challenges, and opportunities in two-dimensional materials beyond graphene. *ACS Nano* 7, 2898–2926 (2013).
[18] Kane, C. L., Mele, E. J., Z_2 topological order and the quantum spin hall effect. *Physical Review Letters* 95, 146802 (2005).
[19] Huang, Y., Pan, Y.-H., Yang, R., et al., Universal mechanical exfoliation of large-area 2D crystals. *Nature Communications* 11, 2453 (2020).
[20] Li, R., Yao, Z., Li, Z., et al., Mechanical exfoliation of non-layered metal oxides into ultrathin flakes. *Nature Synthesis* 4, 106–115 (2025).
[21] Liu, Y., Gradwohl, K.-P., Lu, C.-H., et al., Growth of 28Si quantum well layers for qubits by a hybrid MBE/CVD technique. *ECS Journal of Solid State Science and Technology* 12, 024006 (2023).
[22] Vishwanath, S., Liu, X., Rouvimov, S., et al., Comprehensive structural and optical characterization of MBE grown $MoSe_2$ on graphite, CaF_2 and graphene. *2D Materials.* 2, 024007 (2015).
[23] Sun, L., Yuan, G., Gao, L., et al., Chemical vapour deposition. *Nature Reviews Methods Primers* 1, 5 (2021).
[24] Li, X., Cai, W., An, J., et al., Large-area synthesis of high-quality and uniform graphene films on copper foils. *Science* 324, 1312–1314 (2009).
[25] Paras, N., Yadav, K., Kumar, P., et al., A review on low-dimensional nanomaterials: Nanofabrication, characterization and applications. *Nanomaterials* 13, 160 (2023).
[26] Wolpers, C., Electron microscopy in Berlin 1928–1945. *Advances in Electronics and Electron Physics* 81, 211–229 (1991).
[27] Binnig, G., Rohrer, H., Scanning tunneling microscopy—From birth to adolescence. *Reviews of Modern Physics* 59, 615–625 (1987).
[28] Binnig, G., Rohrer, H., Gerber, C., et al., Surface studies by scanning tunneling microscopy. *Physical Review Letters* 49, 57–61 (1982).
[29] Damascelli, A., Hussain, Z., Shen, Z.-X., Angle-resolved photoemission studies of the cuprate superconductors. *Reviews of Modern Physics* 75, 473–541 (2003).
[30] Ferrari, A. C., Meyer, J. C., Scardaci, V., et al., Raman spectrum of graphene and graphene layers. *Physical Review Letters* 97, 187401 (2006).
[31] Novoselov, K. S., Geim, A. K., Morozov, S. V., et al., Two-dimensional gas of massless Dirac fermions in graphene. *Nature* 438, 197–200 (2005).
[32] Ghosh, C. K., Quantum effect on properties of nanomaterials. In *Introduction to Nano: Basics to Nanoscience and Nanotechnology*, Sengupta, A., Sarkar, C. K., Eds. Springer Berlin Heidelberg, Berlin, Heidelberg, pp. 73–111 (2015).
[33] Edvinsson, T., Optical quantum confinement and photocatalytic properties in two-, one- and zero-dimensional nanostructures. *Royal Society Open Science* 5, 180387 (2018).
[34] Hung, N. T., Hasdeo, E. H., Nugraha, A. R. T., et al., Quantum effects in the thermoelectric power factor of low-dimensional semiconductors. *Physical Review Letters* 117, 036602 (2016).
[35] Trevisanutto, P. E., Giorgetti, C., Reining, L., et al., Ab initio GW many-body effects in graphene. *Physical Review Letters* 101, 226405 (2008).
[36] Cao, Y., Fatemi, V., Fang, S., et al., Unconventional superconductivity in magic-angle graphene superlattices. *Nature* 556, 43–50 (2018).

[37] Ugeda, M. M., Bradley, A. J., Shi, S.-F., et al., Giant bandgap renormalization and excitonic effects in a monolayer transition metal dichalcogenide semiconductor. *Nature Materials* 13, 1091–1095 (2014).
[38] Chernikov, A., Berkelbach, T. C., Hill, H. M., et al., Exciton binding energy and non-hydrogenic Rydberg series in monolayer WS_2. *Physical Review Letters* 113, 076802 (2014).
[39] König, M., Wiedmann, S., Brüne, C., et al., Quantum spin hall insulator state in HgTe quantum wells. *Science* 318, 766–770 (2007).
[40] Novoselov, K. S., Jiang, D., Schedin, F., et al., Two-dimensional atomic crystals. *Proceedings of the National Academy of Sciences* 102, 10451–10453 (2005).
[41] Geim, A. K., Novoselov, K. S., The rise of graphene. *Nature Materials* 6, 183–191 (2007).
[42] Zhu, C., Boehme, S. C., Feld, L. G., et al., Single-photon superradiance in individual caesium lead halide quantum dots. *Nature* 626, 535–541 (2024).
[43] Mourik, V., Zuo, K., Frolov, S. M., et al., Signatures of majorana fermions in hybrid superconductor-semiconductor nanowire devices. *Science* 336, 1003–1007 (2012).
[44] Lin, Q., Fang, H., Kalaboukhov, A., et al., Moiré-engineered light-matter interactions in MoS_2/WSe_2 heterobilayers at room temperature. *Nature Communications* 15, 8762 (2024).
[45] Zou, X., Bai, Y., Dai, Y., et al., Robust second-order topological insulator in 2D van der Waals magnet CrI3. *Materials Horizons* 11, 6416–6422 (2024).
[46] Kezilebieke, S., Huda, M. N., Vaňo, V., et al., Topological superconductivity in a van der Waals heterostructure. *Nature* 588, 424–428 (2020).
[47] Fu, D., Liu, J., Hou, F., et al., Electric field-tunable ferromagnetism in a van der Waals semiconductor up to room temperature. *Nature Communications* 16, 10197 (2025).
[48] Klitzing, K. V., Dorda, G., Pepper, M., New method for high-accuracy determination of the fine-structure constant based on quantized hall resistance. *Physical Review Letters* 45, 494–497 (1980).
[49] Iijima, S., Helical microtubules of graphitic carbon. *Nature* 354, 56–58 (1991).
[50] Novoselov, K. S., Geim, A. K., Morozov, S. V., et al., Electric field effect in atomically thin carbon films. *Science* 306, 666–669 (2004).
[51] Choi, W., Choudhary, N., Han, G. H., et al., Recent development of two-dimensional transition metal dichalcogenides and their applications. *Materials Today* 20, 116–130 (2017).
[52] Bhimanapati, G. R., Lin, Z., Meunier, V., et al., Recent advances in two-dimensional materials beyond graphene. *ACS Nano* 9, 11509–11539 (2015).
[53] Chhowalla, M., Liu, Z., Zhang, H., Two-dimensional transition metal dichalcogenide (TMD) nanosheets. *Chemical Society Reviews* 44, 2584–2586 (2015).
[54] Cho, C., Wong, J., Taqieddin, A., et al., Highly strain-tunable interlayer excitons in MoS_2/WSe_2 heterobilayers. *Nano Letters* 21, 3956–3964 (2021).
[55] Aivazian, G., Gong, Z., Jones, A. M., et al., Magnetic control of valley pseudospin in monolayer WSe_2. *Nature Physics* 11, 148–152 (2015).
[56] Radisavljevic, B., Radenovic, A., Brivio, J., et al., Single-layer MoS_2 transistors. *Nature Nanotechnology* 6, 147–150 (2011).
[57] Castellanos-Gomez, A., Roldán, R., Cappelluti, E., et al., Local strain engineering in atomically thin MoS_2. *Nano Letters* 13, 5361–5366 (2013).
[58] Manzeli, S., Allain, A., Ghadimi, A., et al., Piezoresistivity and strain-induced band gap tuning in atomically thin MoS_2. *Nano Letters* 15, 5330–5335 (2015).
[59] Mak, K. F., He, K., Shan, J., et al., Control of valley polarization in monolayer MoS_2 by optical helicity. *Nature Nanotechnology* 7, 494–498 (2012).

[60] Liu, X., Hersam, M. C., 2D materials for quantum information science. *Nature Reviews Materials* 4, 669–684 (2019).
[61] Dean, C. R., Wang, L., Maher, P., et al., Hofstadter's butterfly and the fractal quantum Hall effect in Moiré superlattices. *Nature*, 497, 598–602 (2013).
[62] Rivera, P., Schaibley, J. R., Jones, A. M., et al., Observation of long-lived interlayer excitons in monolayer $MoSe_2$–WSe_2 heterostructures. *Nature Communications* 6, 6242 (2015).
[63] Wang, J. I. J., Yamoah, M. A., Li, Q., et al., Hexagonal boron nitride as a low-loss dielectric for superconducting quantum circuits and qubits. *Nature Materials* 21, 398–403 (2022).
[64] Gao, X., Vaidya, S., Li, K., et al., Nuclear spin polarization and control in hexagonal boron nitride. *Nature Materials* 21, 1024–1028 (2022).
[65] Mak, K. F., McGill, K. L., Park, J., et al., The valley Hall effect in MoS_2 transistors. *Science* 344, 1489–1492 (2014).
[66] Li, J., Chen, J., He, Y., et al., Controllable growth of monocrystalline MoS_2/polycrystalline ReS_2 lateral heterojunction for high quantum efficiency photodetector. *Advanced Functional Materials* 35, 2421508 (2025).
[67] Liu, S., Wang, X., Xu, N., et al., A flexible and wearable photodetector enabling ultra-broadband imaging from ultraviolet to millimeter-wave regimes. *Advanced Science* 11, 2401631 (2024).
[68] Xu, H., Jiang, Y., Wang, H., et al., Discovering two-dimensional magnetic topological insulators by machine learning. *Physical Review B* 109, 035122 (2024).
[69] Ma, A., Zhang, Y., Christensen, T., et al., Topogivity: A machine-learned chemical rule for discovering topological materials. *Nano Letters* 23, 772–778 (2023).
[70] Gao, X., Vaidya, S., Li, K., et al., Single nuclear spin detection and control in a van der Waals material. *Nature* 643, 943–949 (2025).
[71] Liu, W., Li, S., Guo, N.-J., et al., Experimental observation of spin defects in the van der Waals material GeS_2. *Nano Letters* 25, 16330–16339 (2025).
[72] Switzer, E. D., Reina-Gálvez, J., Giedke, G., et al., Unraveling spin entanglement using quantum gates with scanning tunneling microscopy-driven electron spin resonance. *Nanoscale Advances* 7, 8048–8057 (2025).
[73] Tang, X., Jiang, H., Lin, Z., et al., Wafer-scale vertical 1D GaN nanorods/2D MoS_2/PEDOT: PSS for piezophototronic effect-enhanced self-powered flexible photodetectors. *Nano-Micro Letters* 17, 56 (2024).
[74] Shin, D. H., Jeong, W. U., Choi, S.-H., Strong dual-functional enhancement in flexible photodetection and photovoltaics of two-dimensional large-area WS_2/MoS_2 heterolayers by using a h-BN interlayer. *ACS Applied Materials & Interfaces* 17, 48474–48482 (2025).
[75] Gao, M., Yao, Y., Chen, J., et al., Low-dimensional materials for bioelectronic devices. *Nature Reviews Bioengineering* 4, 171–191 (2026).
[76] Mariello, M., Proctor, C. M., Wireless power and data transfer technologies for flexible bionic and bioelectronic interfaces: Materials and applications. *Advanced Materials Technologies* 10, 2400797 (2025).
[77] Dastgeer, G., Nisar, S., Rasheed, A., et al., Exploring 2D material-based biosensors for real-time detection of target analytes. *ACS Applied Electronic Materials* 7, 7447–7472 (2025).

2 Preparation and Characterization of Low-Dimensional Quantum Materials

Biwen He, Siwei Luo, and Zongyu Huang

2.1 GROWTH TECHNIQUES

The morphology and physical states of low-dimensional quantum materials are directly determined by their fabrication processes. The development and optimization of novel preparation techniques play a critical role in controlling the microstructure and properties of these materials. Since matter in nature primarily exists in gaseous, liquid, and solid phases, mainstream synthesis methods for low-dimensional quantum materials can be categorized into vapor-phase deposition, liquid-phase synthesis, and solid-phase synthesis. These approaches enable precise control over morphology, dimensions, crystal orientation, and composition. The following sections provide detailed discussions of the synthesis protocols for low-dimensional materials across these three categories.

2.1.1 Vapor-Phase Deposition

The history of vapor-phase synthesis of low-dimensional materials dates back to antiquity: the collection of carbon black from candle flames for ink production by ancient civilizations represents a classic example of gas-phase synthesis of nanoparticles. Vapor-phase methods encompass preparation techniques that either utilize gaseous precursors directly or convert reactants into the vapor state by various means, which then undergo physical or chemical transformations in the gas phase to yield low-dimensional materials upon cooling and condensation. Based on whether chemical reactions occur during the process, these methods can be broadly classified into physical vapor deposition (PVD) and chemical vapor deposition (CVD).

PVD, also known as the evaporation–condensation method, involves heating source materials to an atomic vapor state followed by deposition or condensation onto a substrate surface. Specifically, this technique employs physical means to evaporate or sputter material sources into gaseous atoms or molecules under vacuum or low-pressure gas-discharge conditions, thereby forming novel solid coatings with properties distinct from the substrate material.[1] As a vapor-phase condensation

DOI: 10.1201/9781003655305-3

technique, PVD enables the fabrication of low-dimensional quantum materials (Figure 2.1), including nanoparticles with narrow size distributions, nanowires, and nanosheets with refined morphologies, through process parameter optimization.[2–5] Several critical factors govern the product formation in PVD, such as the direction and flow rate of carrier gas (typically inert gases), control deposition location, growth rate, and particle size; substrate temperature modulation (heating or cooling) regulates crystallinity, grain size, and uniformity; and operation within a vacuum system effectively prevents oxidation of both precursors and products during heating. When fabricating metallic and metal oxide nanopowders via PVD, the metal source is evaporated within an ultra-high vacuum chamber filled with a rare gas. Upon reaching supersaturation, the vapor rapidly nucleates and aggregates into clusters, which are then transported by convective flow of the cooled gas to the inner surface of a vertical condensation tube cooled by liquid nitrogen, ultimately forming nanoparticles. One-dimensional (1D) nanostructured quantum materials can be synthesized by evaporating corresponding powdered sources. Taking gallium selenide (GaSe) nanostructure synthesis as an example,[4] the process is conducted in a horizontal tube furnace with polycrystalline GaSe powder placed at the center as the source and a silicon oxide substrate positioned downstream as the deposition target. Prior to heating, the furnace is evacuated to high vacuum and repeatedly purged with a carrier gas to eliminate residual oxygen. The source powder is then heated to its sublimation temperature, and the carrier gas transports vapor molecules to the substrate surface where product growth occurs.

PVD exhibits exceptional versatility, enabling uniform growth of high-quality two-dimensional (2D) thin films over large-area substrates. This technique effectively controls film composition, facilitating the fabrication of heterostructures and layered materials required for quantum sensor devices. PVD has been successfully employed for the growth of 2D material heterostructures. For instance, Haoxin Huang et al. utilized PVD to fabricate vertically stacked 2D heterojunctions and applied them in quantum sensors.[5] The capability to engineer such structures via PVD enhances the sensitivity and precision of quantum state detection, which is critical for quantum sensing applications. In contrast to the aforementioned evaporation process, in which products form via gas-phase particle condensation, sputtering involves ejecting target atoms or molecules from a solid surface under bombardment by high-energy particles. Sputtered particles exhibit higher energies compared to evaporated species, along with pronounced directionality and superior film-forming characteristics. As a variant of PVD, sputtering can be classified based on ion excitation and film deposition mechanisms into direct current sputtering, ion beam sputtering, radio frequency sputtering, and magnetron sputtering. Magnetron sputtering, which has developed rapidly since the 1970s, is a high-rate deposition technique. This method introduces a magnetic field near the cathode target, utilizing the Lorentz force to extend electron trajectories in the electric field, thereby increasing the probability of collisions between electrons and gas molecules and enhancing ionization efficiency.[6] As shown in Figure 2.2, this process substantially augments ion flux bombarding the cathode target, consequently improving sputtering rates. Under optimized process conditions, magnetron sputtering has achieved remarkable success in fabricating metallic cluster nanoparticles, with particularly extensive applications in

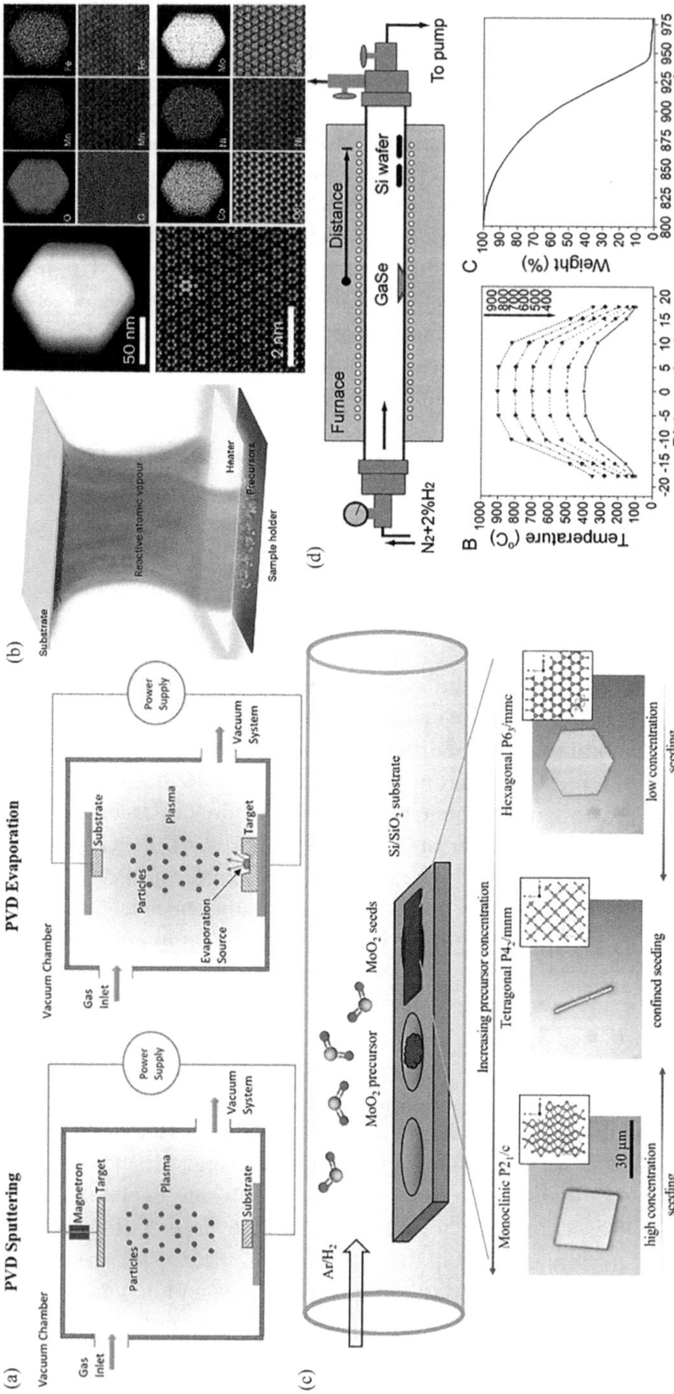

FIGURE 2.1 (a) Schematic illustration of the PVD process.[1] (b)–(d) Representative examples of low-dimensional quantum materials fabricated by PVD.[2–4]

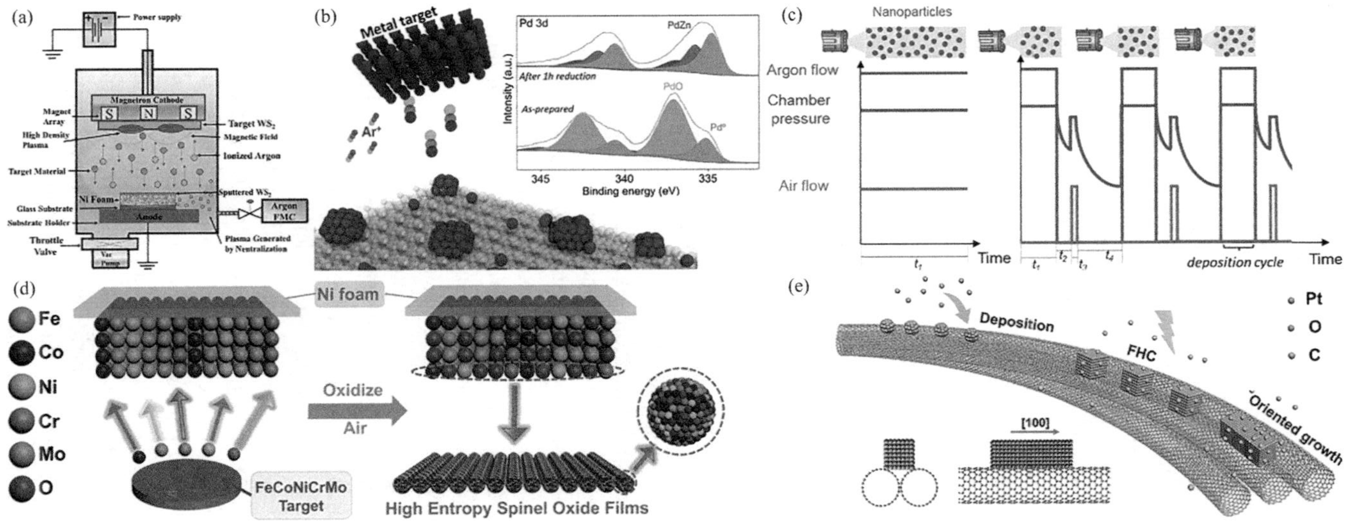

FIGURE 2.2 (a) Schematic diagram of magnetron sputtering.[6] (b)–(e) Representative examples of nanoparticles and thin films fabricated by magnetron sputtering.[7–10]

the synthesis of alloy nanoparticles.[7–9] Direct sputtering from alloy targets yields nanoparticles with compositions identical to those of the target materials without requiring additional control measures.

CVD is the dominant vapor-phase growth technique for low-dimensional materials, enabling the fabrication of quantum materials across multiple dimensions, including nanowires and nanofilms. Through precise control of process parameters, this method yields both high-quality pristine samples with simple structures and complex architectures incorporating screw dislocations and helical morphologies.[11,12] Extensive mechanistic studies of CVD growth have been conducted through theoretical modeling and in situ characterization, with the vapor–liquid–solid (V–L–S) and vapor–solid–solid (V–S–S) mechanisms being the most extensively validated. The V–L–S mechanism was first proposed by Wagner and Ellis at Bell Laboratories in the 1960s during investigations into the synthesis of high-quality silicon carbide whiskers.[13] Its essential process involves catalytic decomposition of gaseous reactants in the presence of a catalyst, forming eutectic liquid droplets; when reactant concentration reaches supersaturation, precipitation and sustained growth occur at the solid–liquid interface. In contrast, V–S–S-mediated growth proceeds entirely in the solid state without liquid-phase participation. Related solid-catalyst growth behavior was directly visualized by Helveg et al. using time-resolved in situ high-resolution transmission electron microscopy during carbon nanofibre growth on nickel nanocrystals; the nickel particles remained crystalline and well faceted throughout growth.[14]

CVD technology (Figure 2.3) enables the synthesis not only of 1D quantum materials, including semiconductor nanowires, nanotubes, and quantum dots, but has also evolved into a mainstream fabrication approach for 2D quantum materials.[15–17] Following the initial demonstration of monolayer graphene growth on copper foil substrates,[18] this technique has been successfully applied to produce diverse 2D quantum materials, including graphene, hexagonal boron nitride (h-BN), transition metal dichalcogenides (TMDC, with general formula: MX_2), ternary compounds, and van der Waals heterostructures.[19–24] Based on variations in reaction energy–supply mechanisms, CVD can be categorized into distinct types for experimental differentiation and process selection, encompassing atmospheric pressure chemical vapor deposition, low-pressure chemical vapor deposition, thermal chemical vapor deposition, laser chemical vapor deposition, metal–organic chemical vapor deposition, and plasma-enhanced chemical vapor deposition. Matching energy–supply mechanisms to the specific characteristics of different reactants further enhances sample-preparation efficiency. In graphene synthesis, CVD growth typically employs pure methane as the gaseous precursor, with controlled gas flow rates that enable reproducible growth. In contrast, MX_2 growth commonly utilizes dual solid-source precursors: metal sources (metal foils, metal oxides, and metal chlorides) and non-metal sources (S, Se, and Te) are positioned in separate temperature zones, vaporized upon heating, and transported by carrier gas to the high-temperature reaction zone where chemical reaction occurs on the substrate surface to form MX_2. The vapor pressure of solid precursors is determined solely by heating temperature; precise control of vaporization rates through temperature modulation alone remains challenging, complicating precise control over MX_2 growth. To address this limitation,

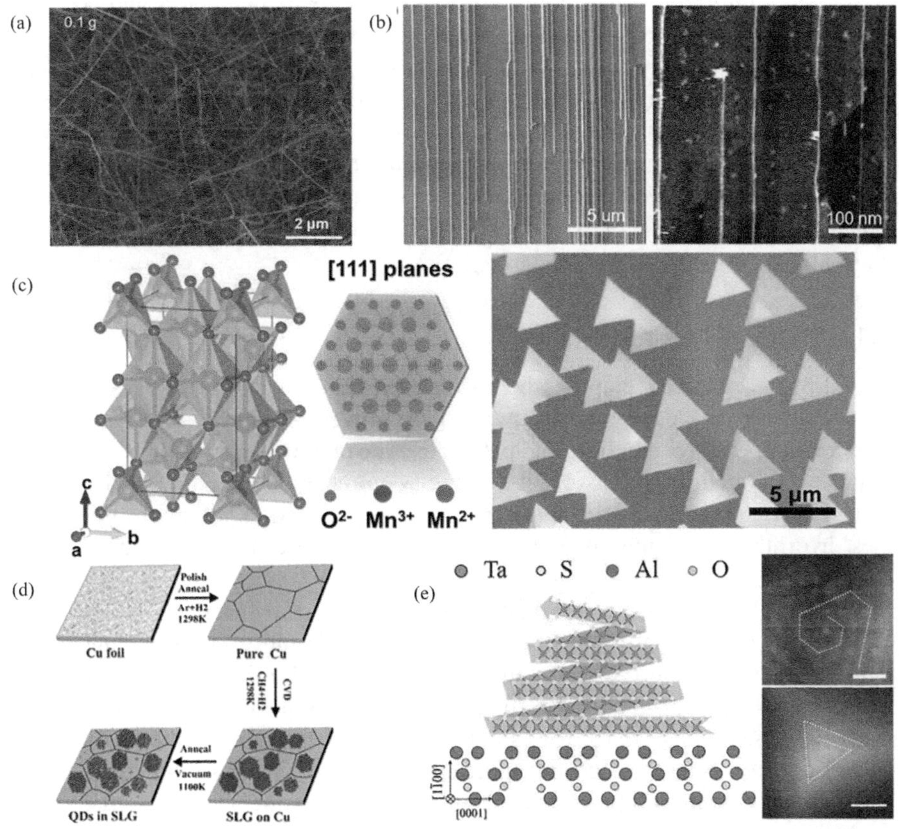

FIGURE 2.3 Representative examples of low-dimensional quantum materials fabricated by CVD.[11,12,15–17]

metal–organic precursors demonstrate significant advantages in the controlled synthesis of 2D materials, enabling accurate and stable precursor delivery within CVD systems. In this regard, Kang et al. successfully synthesized continuous, 4-inch wafer-scale MoS_2 and WS_2 films using $Mo(CO)_6$ and $W(CO)_6$ as metal precursors, respectively, and diethyl sulfide $[(C_2H_5)_2S]$ as the non-metal precursor.[21]

Furthermore, in practical CVD experiments, researchers frequently optimize conventional protocols by addressing existing instrumental limitations, thereby developing novel experimental methodologies. In standard CVD systems, product thickness and domain size are controlled solely by adjusting growth temperature, carrier gas, precursor flow rates, and chamber pressure. However, complex intermediate reaction processes inevitably introduce issues including thickness non-uniformity, impurity adsorption, and sluggish growth kinetics. To improve material uniformity and enhance growth efficiency, space-confined growth has emerged as a promising optimization strategy. This structural design accelerates the decomposition of methane precursors and reduces nucleation density, thereby enabling rapid fabrication of large-area graphene single crystals.[25] Additionally, to address technical challenges associated with high melting points of metal/metal oxide precursors and poor thermodynamic stability of metastable phases, researchers have developed modified CVD processes employing molten salts as fluxes, successfully synthesizing diverse high-quality 2D materials.[26] Mechanistic studies reveal that salt additives reduce reactant melting points and facilitate formation of oxychloride intermediates, which are energetically favorable for subsequent sulfidation and selenization reactions. Concurrently, alkali metal ions effectively lower growth barriers and synergistically enhance overall reaction rates, providing viable pathways for scalable fabrication of high-quality, uniform 2D materials.

2.1.2 Liquid-Phase Synthesis

Liquid-phase synthesis employs homogeneous solutions as starting systems, with chemical reactions controlled by modulating solvent composition, concentration, reaction temperature, pressure, and pH to generate target products. The solute is subsequently isolated from the solvent to serve as a precursor, then dried and thermally treated to yield the desired nanomaterials. Among diverse liquid-phase methodologies, hydrothermal synthesis and sol–gel processing are the most widely utilized categories.

The term "hydrothermal" originated approximately 150 years ago in the geological sciences, initially describing the natural evolution of subsurface water under combined temperature and pressure conditions within the Earth's crust. Subsequently, this terminology was widely adopted across various chemical processes to characterize specific reaction systems. It was not until the 1970s that hydrothermal synthesis was recognized as an advanced technique for fabricating inorganic nanoparticles. As a significant branch of liquid-phase chemistry, hydrothermal synthesis (also termed the hydrothermal method) refers to the preparation of nanomaterials under high-pressure conditions within sealed high-pressure reactors (autoclaves), utilizing aqueous solutions as the reaction media. Heating creates elevated temperature and pressure conditions that enable the dissolution and recrystallization of substances that are normally insoluble or poorly soluble, thereby facilitating nanomaterial synthesis. Recent works (Figure 2.4) demonstrate that hydrothermal synthesis is a prevalent wet-chemical method for preparing high-quality quantum dots, including graphene/reduced graphene oxide, molybdenum disulfide (MoS_2), and tungsten disulfide (WS_2). This method exfoliates bulk three-dimensional (3D) layered precursors into small fragments by breaking van der Waals and covalent bonds in high-temperature, high-pressure aqueous environments.[30] Additionally, regarding TMDC materials, research groups have employed hydrothermal methods to synthesize WS_2 structures with various morphologies by introducing surfactants such as cetyltrimethylammonium bromide and polyethylene glycol (PEG).[28] This method is particularly applicable to reactions unfeasible under ambient conditions, successfully producing various oxide nanoparticles and limited water-insensitive sulfides, including representative examples such as ZrO_2, Al_2O_3, TiO_2, and CrO_2.[29] Since hydrothermal synthesis typically operates below 300°C, it is especially suitable for materials prone to decomposition or destabilization at elevated temperatures. However, the intrinsic properties of water as a reaction medium also impose limitations.

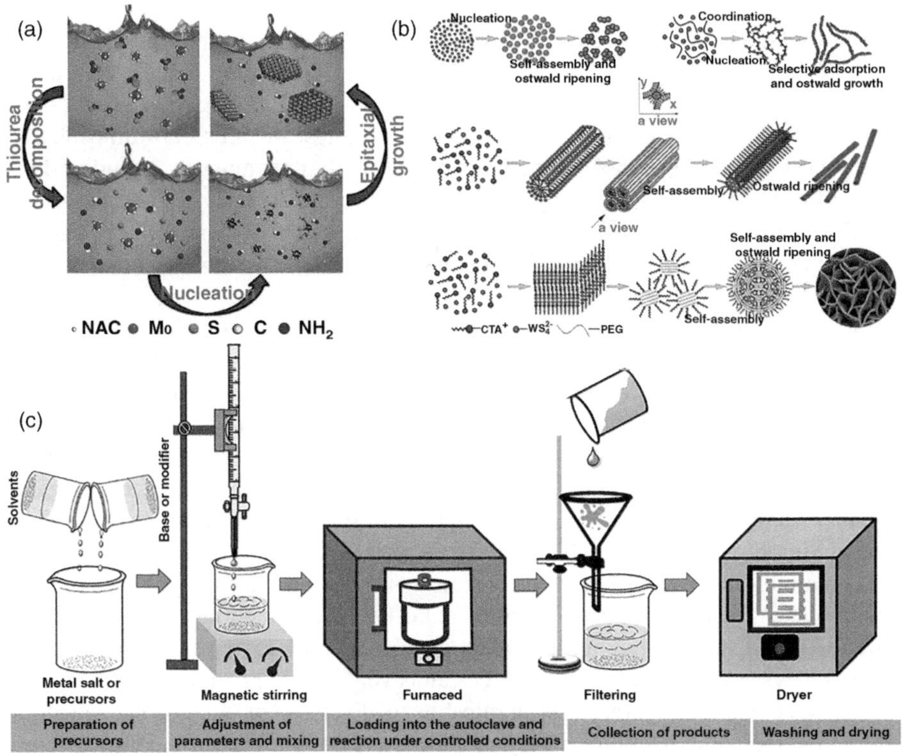

FIGURE 2.4 (a) Schematic illustration of the hydrothermal method.[27] (b) Representative example of low-dimensional quantum materials synthesized by the hydrothermal method.[28] (c) Experimental flowchart of the hydrothermal method.[29]

Hydrothermal methods are incompatible with water-sensitive materials and systems in which reactants or products undergo hydrolysis or oxidation, thereby constraining the scope of applicable synthetic targets. Solvothermal synthesis represents an advanced liquid-phase technique evolved from hydrothermal methods. This approach employs organic solvents, including alcohols, benzene, ammonia, and organic amines, as reaction media in place of water, enabling fabrication of materials susceptible to hydrolysis or oxidation that are unattainable in aqueous systems. Within solvothermal reaction environments, solvents function not merely as dissolution and pressure-transmission media with mineralizing effects, but also participate directly in chemical reactions to regulate product crystalline phases and morphologies. Consequently, the selection of appropriate organic solvents constitutes the critical step in solvothermal synthesis, requiring a comprehensive evaluation of the physicochemical properties of water/organic solvent systems. For example, sulfur quantum dots exhibiting multiple fluorescence emissions and distinct size distributions can be synthesized by ultrasonically treating sulfur powder, PEG-400, and hydrogen peroxide in ethanol, followed by heating at 220°C. The use of hydrogen peroxide as both an etchant and a solvent further enhances the fluorescence quantum yield and reduces the reaction time.[31] The fabrication process for 2D films via hydrothermal methods proceeds entirely within liquid-phase systems without requiring subsequent crystallization heat treatment, effectively preventing defects commonly associated with thermal processing, including film curling, cracking, grain coarsening, and reactions between films and substrates or ambient atmospheres. Hydrothermal and solvothermal methods typically offer advantages, including high crystallinity, minimal particle agglomeration, high product purity, and narrow particle-size distributions. These characteristics render them environmentally benign, cost-effective, commercially viable, and highly competitive approaches for low-dimensional material synthesis.

A sol is a colloidal system with liquid-like characteristics, containing solid particles 1–100 nm in diameter suspended and dispersed within a liquid medium and undergoing random Brownian motion. A gel is a colloidal system with solid-like properties, in which dispersed colloidal particles interconnect to form a continuous network skeleton with liquid- or gas-filled interstitial pores. Sol–gel synthesis of low-dimensional quantum materials employs solutions of metal–organic or inorganic compounds as starting precursors. Through hydrolysis and polycondensation reactions in liquid-phase systems (Figure 2.5), the original solution is transformed into a sol containing metal oxide particles; subsequent gelation yields a gel, which, upon heating and drying, produces various nanostructured materials.[32] This method requires homogeneous mixing and reaction of reactants in solution, formation of stable sol systems from reaction products, and avoidance of precipitation throughout the process. Sol–gel synthesis is applicable to diverse low-dimensional quantum structures, including quantum dots, nanopowders, and thin films.[35,36] Three primary technical routes exist for 2D film fabrication, including dip coating, spin coating, and laminar flow coating.[34] Each method possesses distinct technical characteristics; selection depends on substrate dimensions, geometry, and target film performance requirements, with dip coating and spin coating being more extensively applied. Precise control of temperature, chemical reagent selection, and reactant concentration effectively modulates the morphology, dimensions, and uniformity of solution-phase-grown nanostructures. For instance, metal alkoxides that undergo facile hydrolysis with pH-tunable reaction kinetics are preferred as precursors; identical alkoxides exhibit distinct polycondensation pathways that yield distinct morphologies and structures under different pH conditions. Additionally, alkoxide addition rate and reaction temperature directly influence gel formation efficiency and structural homogeneity. Compared with alternative preparation methods, sol–gel synthesis exhibits exceptional versatility, enabling fabrication of large-area films on diverse substrate geometries, including planar, cylindrical, tubular inner surfaces, spherical, and fibrous configurations, and various materials such as metals, glass, ceramics, and polymers, as well as coating formation on powder surfaces.[33] Reports indicate that TiO_2 films deposited on glass and silica substrates via sol–gel methods exhibit varying morphologies and properties depending on the substrate.[37] Mengdi Hu et al. fabricated indium tin oxide (ITO) films on silicon wafers, demonstrating a spectral responsivity of 5.6 A/W at 450 nm and strong laser absorption.[38] Furthermore, precursor solutions of target components can be homogeneously mixed in sol–gel processes, facilitating quantitative doping and precise control over powder particle size during fabrication.

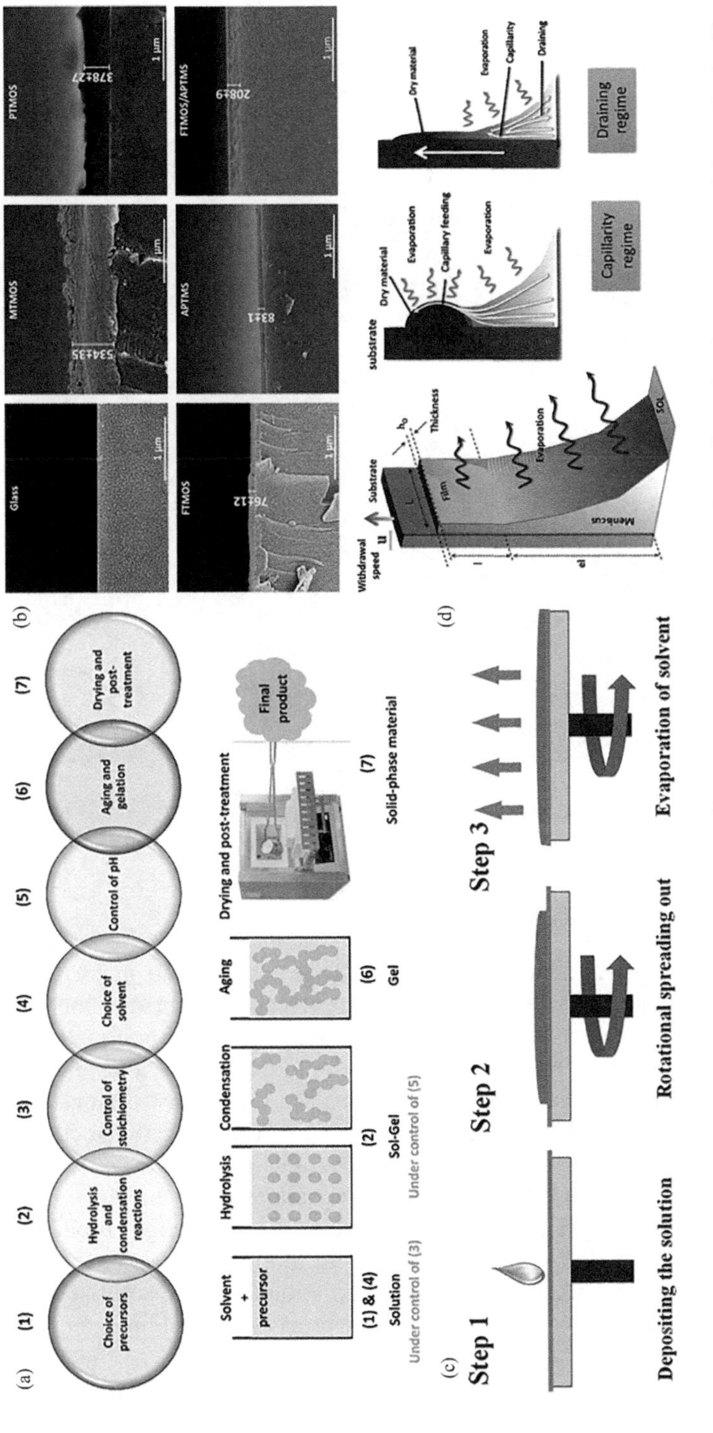

FIGURE 2.5 (a) Schematic illustration of the sol–gel method.[32] (b) Representative example of low-dimensional quantum materials synthesized by the sol–gel method.[33] (c) and (d) Schematic diagrams of spin coating and dip coating, respectively.[34]

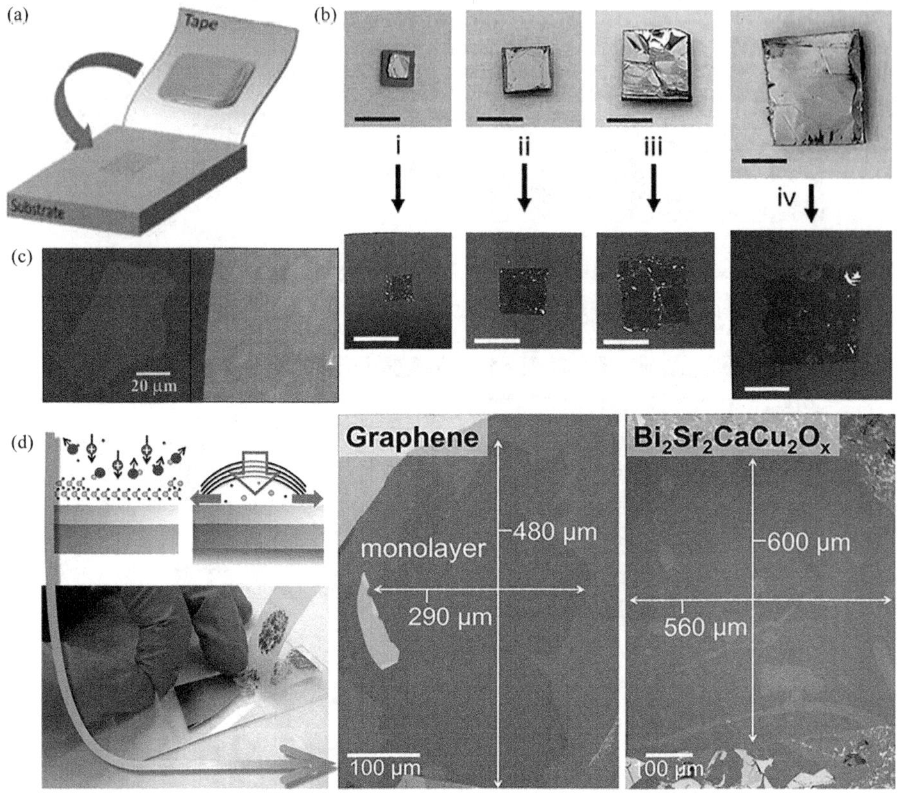

FIGURE 2.6 (a) Schematic illustration of mechanical exfoliation.[40] (b)–(d) Representative examples of low-dimensional quantum materials prepared by mechanical exfoliation.[39,41,42]

2.1.3 Solid-Phase Synthesis

Solid-phase synthesis encompasses methods in which solid materials serve as the principal reactants or starting materials. Depending on the mechanism, it includes top-down mechanical methods and solid-state reaction methods. Solid-state methods offer technical advantages, including high yield and simplified processing workflows, and remain widely employed for low-dimensional quantum material fabrication. Based on mechanistic distinctions, this methodology is primarily classified into mechanical methods and solid-state reaction methods. Mechanical methods directly utilize starting materials, achieving particle-size reduction through physical mechanical action to obtain target products. Solid-state reaction methods require stoichiometric mixing of reactants, followed by uniform grinding and high-temperature calcination to induce solid-state reactions that yield direct product, with subsequent processing yielding the final target products.

Mechanical exfoliation, commonly known as the scotch tape method, represents one of the earliest and most straightforward techniques for fabricating high-quality two-dimensional materials, first proposed and demonstrated by Novoselov and colleagues during graphene preparation.[39] This method has been extended beyond graphite to various inorganic layered materials (Figure 2.6).[40] The underlying principle involves application of shear forces to bulk crystals to induce interlayer exfoliation, yielding 2D material flakes. The exfoliated films are typically transferred onto substrates such as silicon wafers for subsequent characterization and device fabrication. Conventional tape-assisted exfoliation exhibits notable limitations: the obtained monolayer 2D materials are relatively small in size (approximately 5 μm), preparation yields remain low, and layer-selective exfoliation is unattainable. To address these limitations of mechanical exfoliation, research groups have proposed several refinement strategies, notably gold-assisted exfoliation and novel oxygen plasma-assisted exfoliation. Gold-assisted exfoliation leverages the bonding interaction between gold films and chalcogen atoms: the Au-TMDC interfacial interaction is stronger than the interlayer van der Waals interaction, enabling efficient exfoliation of large-area monolayer samples.[43] Photoluminescence (PL) spectroscopy characterization demonstrates that ultra-large monolayer samples obtained via gold-mediated exfoliation exhibit comparable optoelectronic performance to tape-exfoliated flakes without degradation.[41]

Oxygen plasma-assisted exfoliation introduces a pretreatment step prior to conventional mechanical exfoliation: substrate surfaces are first cleaned with oxygen plasma to remove contaminants, then subjected to additional thermal treatment. This pretreatment effectively enhances adhesion between crystals and substrates, thereby improving the production yield of graphene nanosheets and achieving efficient graphene exfoliation. Graphene nanosheets fabricated through these improved methods have been successfully integrated into graphene-based field-effect transistors, with subsequent carrier mobility measurements confirming excellent electrical performance and validating the efficacy of these refinements.[42] The distinctive advantage of mechanical exfoliation lies in its capability for controlled exfoliation of monolayer or few-layer materials, a critical attribute for quantum sensing applications where intrinsic material properties must be fully preserved. This technique yields monolayer or few-layer materials with minimal defects, making them ideal candidates for sensors that operate on quantum effects. However, as graphene research demonstrates, this method exhibits significant limitations: monolayers obtained through micromechanical exfoliation constitute a minute fraction among flakes of various thicknesses, precluding scalable fabrication. Consequently, although this method enabled the initial characterization study of monolayer materials, it does not constitute a viable large-scale preparation process for industrial applications of 2D materials.

Compared with graphene, h-BN possesses stronger chemical bonding, rendering effective preparation via mechanical exfoliation challenging. Accordingly, ball milling has been established as an effective approach for scalable fabrication of h-BN nanosheets, offering practical value despite certain inherent limitations.[44] Ball milling employs mechanical grinding to exfoliate bulk materials into thin films, with optimized processes that preserve large-area single-crystalline domains. Beyond 2D film preparation, ball milling is extensively used in nanoparticle synthesis.[45] High-energy ball milling (also termed mechanochemistry) generates new substances through direct mechanical energy participation or initiation of chemical reactions.

The underlying principle involves the utilization of rotational or vibrational motion in milling equipment to induce intense impact, grinding, and mixing of hard balls against raw materials; during milling, samples undergo repeated deformation leading to critical defect density in localized strain bands, causing grain fracture. This process iterates until grain refinement reaches the nanoscale. High-energy ball milling enables convenient preparation of metallic nanoparticles with cubic and hexagonal close-packed structures.

In the narrow sense, solid-state reactions specifically denote chemical reactions occurring between solid phases to generate novel solid products; in the broad sense, they encompass all chemical reactions involving solid phases, including solid decomposition, oxidation, and reactions between solid–solid or solid–liquid interfaces. As early as 2010, research groups successfully fabricated $SrFe_2As_{2-x}P_x$ superconductors via solid-state reaction methods. The preparation protocol involved grinding and pelletizing metallic strontium chunks, arsenic granules, iron powder, and red phosphorus as starting materials, followed by sealing in quartz ampoules and sintering. Upon cooling to room temperature, the target product was obtained, which exhibited excellent superconducting properties upon characterization.[46]

The reaction efficiency and product structure of solid-state reactions are directly governed by the chemical composition, crystal structure, and particle size of reactants. This arises because solid-state reactions are heterogeneous processes that predominantly occur at interfaces between different components, with effective contact between reactants being a prerequisite for chemical reaction and mass transport. The most prominent advantage of solid-state reactions lies in their fundamental adoption of pollution-prevention strategies at the source, substantially reducing or eliminating contaminants; the solvent-free nature of this method intrinsically minimizes pollution generation, thereby effectively preventing environmental contamination.

2.2 CHARACTERIZATION TECHNIQUES

2.2.1 Microstructural and Morphological Characterization

Imaging techniques constitute core methodologies for obtaining direct visual information in materials characterization. Commonly employed methods include optical microscopy, X-ray imaging, scanning electron microscopy (SEM), transmission electron microscopy (TEM), scanning tunneling microscopy (STM), and atomic force microscopy, enabling precise acquisition of critical information regarding surface morphology, crystal structure, atomic arrangement, pore architecture, defect states, and interfacial characteristics (Figure 2.7).

Optical microscopy operates on the principle of magnified imaging through convex lenses, enlarging morphological features of samples at millimeter to micrometer scales to facilitate preliminary analysis. As the most fundamental tool in materials characterization, optical microscopy offers operational simplicity and high detection efficiency, enabling convenient assessment of growth uniformity, dimensions, and grain boundaries. Its non-destructive nature supports exploration of material growth conditions and optimization of device fabrication processes. For instance, under optical microscopy, 2D material films of varying thicknesses exhibit distinct color

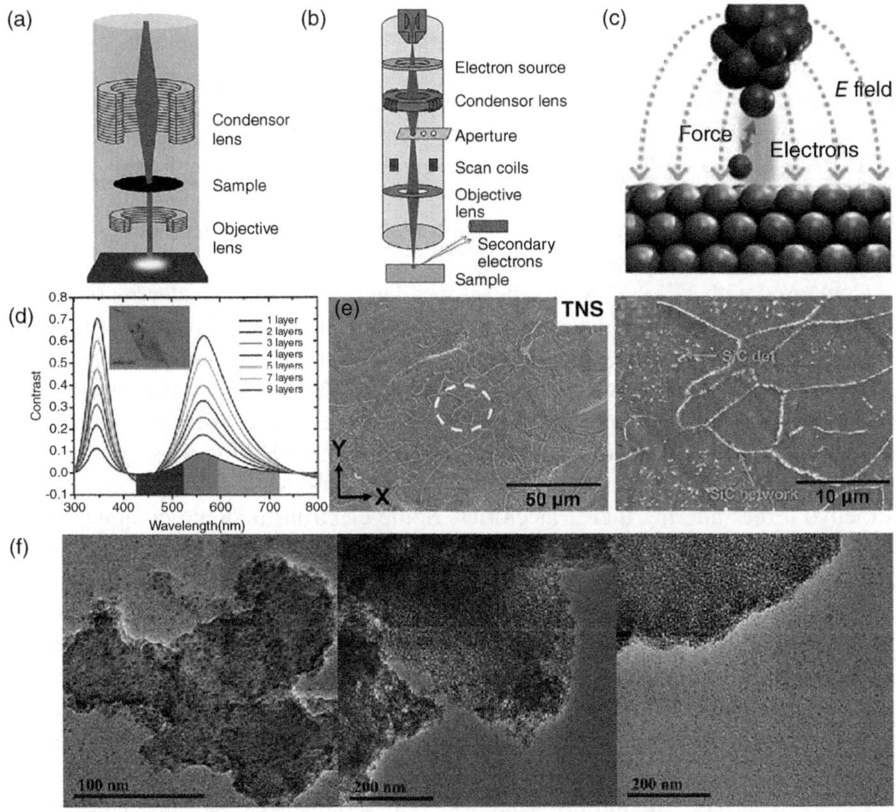

FIGURE 2.7 (a) Schematic diagram of TEM.[47] (b) Schematic diagram of SEM.[48] (c) Schematic illustration of atomic manipulation by STM.[49] (d) Optical contrast versus layer number of the sample under optical microscopy.[50] (e) SEM image of the sample.[51] (f) TEM image of the sample.[52]

characteristics: samples with coloration closer to the substrate indicate fewer layers, whereas pronounced color contrast signifies greater thickness.[50,53] Conventional optical microscopy is constrained by optical physics, with magnification limited to 500× or 1000× and minimum resolvable dimensions of approximately 0.2 μm. This limitation was recognized as early as the 1930s, prompting efforts to overcome this technical bottleneck. TEM represents the first electron microscope developed. Electron microscopes employ high-energy electron beams to detect minute objects, with high-resolution instruments achieving spatial resolution well below 1 nm, reaching the sub-angstrom regime in modern aberration-corrected instruments. These systems simultaneously provide structural information in both real and reciprocal spaces, as well as chemical composition analysis. TEM examines electron-transparent nanomaterials, utilizing focused electron beams as illumination sources and transmitted electrons as imaging signals; the resulting images reflect 2D structural characteristics.[47] For example, Sarah K et al. directly visualized the morphology and particle size of silicon quantum dots via TEM, confirming the formation of 2–5 nm high-purity, low-oxidation quantum dots, providing critical morphological and compositional evidence for the successful preparation of energetic silicon quantum dots.[52] For suitable samples, three-dimensional information can be obtained by electron tomography using a tilt series of TEM images. Recent breakthroughs in aberration correction, in situ experimental platforms, and ultrafast detection techniques have established TEM as an irreplaceable tool in frontier research domains, including high-temperature superconductivity, topological quantum states, low-dimensional materials, and defect engineering, thereby making it a central characterization methodology in low-dimensional quantum materials research.[54,55]

Compared with TEM, SEM was developed later but enjoys broader applicability. SEM serves as an intermediate microscopic observation technique between TEM and optical microscopy, operating on the principle of scanning a sample surface with a focused, narrow, and high-energy electron beam. Various physical signals generated through electron-sample interactions are collected, amplified, and imaged to enable precise characterization of surface microstructure. During SEM observation, incident electrons interact with atomic nuclei via elastic Rutherford scattering, with attractive forces altering their trajectories. Some electrons undergo complete backscattering and escape from the sample surface, with scattering angles closely correlated with atomic number. Consequently, images formed from detected backscattered electrons characterize sample geometry and atomic composition.[51] SEM offers distinct technical advantages, including exceptionally high resolution typically reaching nanometer scales; substantial depth of field enabling clear 3D topographic presentation unmatched by other electron microscopes; simplified sample preparation; capability for large-area imaging; and an exceptionally wide magnification range ($10–10^5\times$).

STM is a surface structural research tool with atomic-scale resolution, achieving sub-atomic-diameter resolution, for precise surface scanning and detailed topographic mapping. It was the first instrument to achieve clear atomic-scale imaging of sample surfaces. STM operates on quantum tunneling principles: tunneling current generated between the sample and the probe tip through quantum mechanical effects exhibits a strong dependence on tip-sample separation, with variations in current reflecting surface electronic corrugation. Atomic-scale observation and manipulation require piezoelectric actuators and probe tips. Quantum tunneling requires conductive or semiconducting properties in both the sample and the tip. STM effectively overcomes inherent limitations of conventional electron microscopes, including radiation damage from high-energy electrons, limited resolution of surface corrugation, and vacuum–environment requirements, enabling non-destructive observation of surface structures in ultrahigh vacuum, atmospheric, and even liquid environments with authentic 3D topographic representation. The most prominent advantage of STM lies in its ability to manipulate and arrange individual atoms and molecules to form novel substances or fabricate materials with specific quantum properties, which is of significant importance for the research and assembly of low-dimensional quantum materials.[49] Although high-resolution electron microscopy can theoretically obtain fringe images, lattice images, structure images, and atomic images of crystals, acquiring structure images, atomic images, and fine intra-atomic structural details remains challenging and requires complementary characterization tools.

2.2.2 Chemical Composition and Crystallographic Analysis

Chemical composition and crystalline phase significantly influence the fabrication process and physicochemical properties of low-dimensional quantum materials, constituting fundamental determinants of their application characteristics. Consequently, systematic characterization of chemical constituent types and concentrations (particularly trace additives), impurity levels, and distribution characteristics, and crystalline phases is essential and indispensable to low-dimensional quantum material research. The chemical composition encompasses major constituents, minor components, additives, and impurities; crystalline phases include crystallinity, crystal structure types, defects, and dislocations. Depending on specific characterization requirements, as shown in Figure 2.8, techniques such as characteristic X-ray analysis, X-ray diffraction (XRD), X-ray photoelectron spectroscopy (XPS), and mass spectrometry enable the precise determination of chemical composition and crystalline state.

Characteristic X-ray analysis operates on the principle of irradiating target microregions on sample surfaces with electron beams; high-energy electrons excite constituent elements, generate characteristic X-rays with specific wavelengths (or energies). Distinct elements produce unique characteristic X-rays, enabling qualitative analysis based on the wavelengths (or energies) of the emitted radiation and quantitative analysis through radiation intensity measurements. This technique integrates microanalysis with compositional analysis for micro-regional characterization, particularly suited to chemical composition and crystalline-state analyses of minute sample areas. In doping engineering studies, characteristic X-ray analysis has been employed to precisely determine the elemental composition and atomic doping ratio of Sn-doped α-Fe_2O_3 films, verifying successful doping and supporting mechanistic analyses of lattice distortion and performance modulation.[58] In contrast, XRD precisely reveals atomic periodicity through analysis of material responses to XRD, playing an irreplaceable role in distinguishing crystalline from amorphous substances. XRD patterns enable qualitative phase identification by comparison with standard reference patterns and quantitative analysis of phase composition, crystallinity, and grain size through diffraction intensity analysis. Dong Hyun Kim et al. employed in situ high-temperature XRD to monitor the temperature-dependent evolution of peak position, intensity, and broadening in barium titanate (BTO) powder during high-temperature phase transitions. They observed that BTO undergoes a tetragonal-to-cubic phase transition at approximately 128°C, accompanied by significant peak broadening near the transition point.[58]

XPS is a surface analysis technique that uses X-ray irradiation of sample surfaces and detection of the kinetic energies of photoemitted electrons to precisely determine surface elemental composition and chemical states. The underlying principle derives from the photoelectric effect, in which X-rays of energy $h\nu$ irradiate sample surfaces, causing inner-shell electrons in constituent atoms to be emitted as photoelectrons. Instruments measure photoelectron kinetic energies to deduce original binding energies. Since binding energies vary significantly across different elements, electron orbitals, and chemical environments, XPS enables precise identification of surface elemental species and chemical states. Furthermore, XPS exhibits exceptional surface sensitivity with

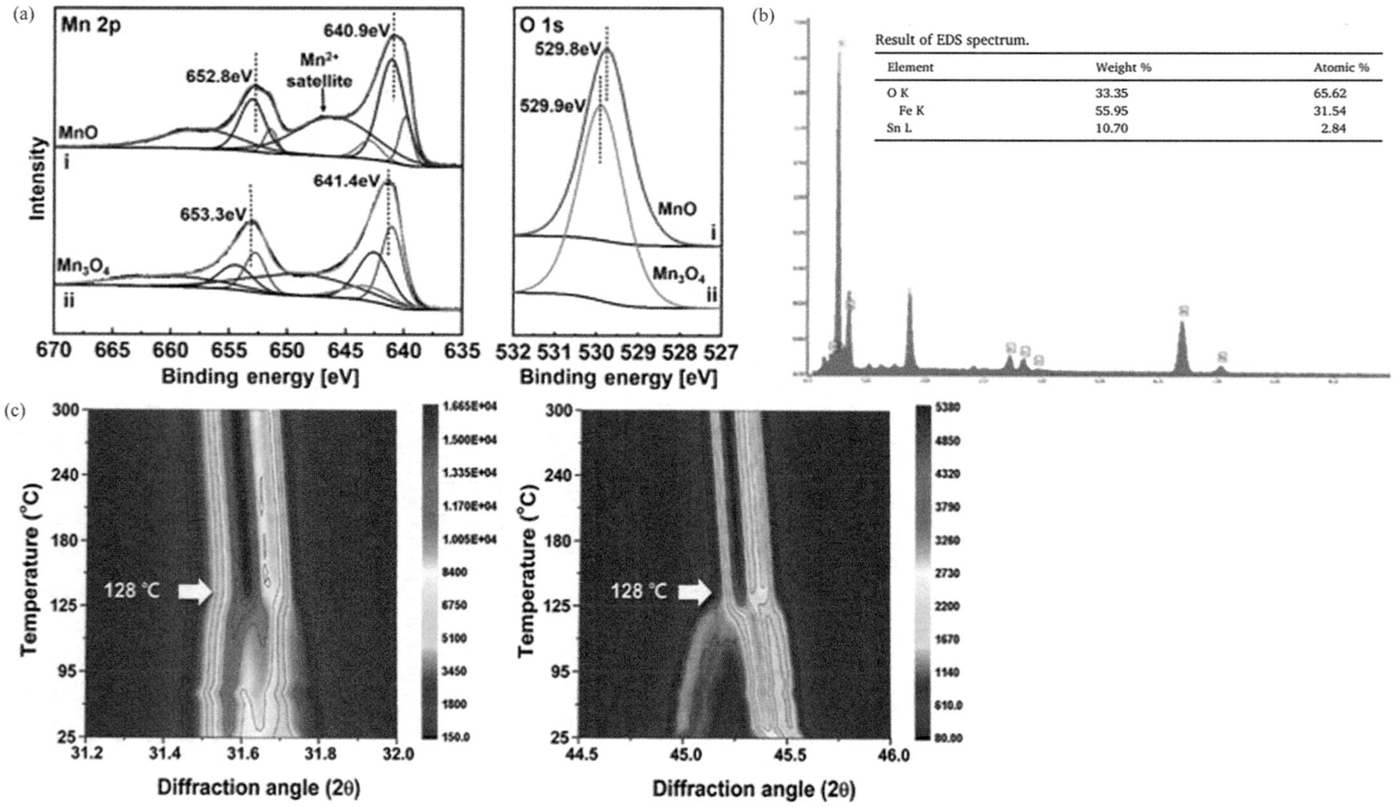

FIGURE 2.8 (a) XPS spectrum of the sample.[56] (b) Energy-dispersive X-ray spectroscopy (EDS) analysis of the sample.[57] (c) XRD pattern of the sample.[58]

detection depths commensurate with those of low-dimensional quantum materials, effectively eliminating substrate signal interference. This capability permits precise characterization of surface elemental composition, chemical valence states, surface states, and interfacial interactions in low-dimensional quantum materials, with particular applicability to 2D materials, quantum dots, and related low-dimensional quantum systems. The literature demonstrates the use of XPS to determine the phase, Mn oxidation state, and surface chemical composition of Mn_3O_4 films.[56]

2.2.3 Performance Characterization

Quantum effects are the most fundamental physical phenomena in quantum materials—a key factor in understanding their exotic physical properties and behaviors. These effects not only elucidate microscopic operating mechanisms but also significantly influence optical properties, electrical conductivity, magnetism, and other performance characteristics of low-dimensional quantum materials. The following sections introduce several typical characterization methods for these properties.

Optical spectroscopy, leveraging its non-invasive nature, high sensitivity, ultrafast temporal resolution, and nanoscale spatial resolution, has become an irreplaceable experimental tool for exploring low-dimensional quantum materials. PL refers to the physical phenomenon wherein a substance absorbs energy from an external light source to become excited, subsequently emitting energy in the form of light radiation. In essence, optical excitation promotes electrons to excited states, followed by radiative relaxation to lower-energy states with photon emission. Overall, the PL process comprises light absorption, energy transfer, and light emission, with interlevel transitions occurring during the first and third stages, while the second stage involves particle motion within the excited state. Under 2D confinement, the dramatically weakened dielectric screening leads to significantly enhanced Coulomb interactions, imparting excitons (electron-hole pairs) with exceptionally high binding energies that enable their stable existence at room temperature and dominant contribution to optical responses. PL spectroscopy not only elucidates bandgap evolution in low-dimensional quantum materials but also captures fine quantum many-body behaviors, including charged excitons, biexcitons, dark excitons, and defect-bound states. The unique quantum states in low-dimensional quantum materials provide stable electron reservoirs with exceptionally high electrical conductivity, carrier mobility, and long-lived spin states. These characteristics (Figure 2.9) not only simplify the understanding of surface responses but also facilitate the exploration of high-performance materials.

Nengjie Huo et al. employed PL mapping to directly compare the emission intensities of various TMDC materials on FePt substrates, confirming that interfacial magnetization significantly enhances the PL intensity of monolayer WS_2.[59] Additionally, Gautham Kumar et al. utilized PL spectroscopy to characterize the emission intensity, quantum yield, and fluorescence lifetime of perovskite quantum dots, revealing the mechanism by which gold nanoparticles accelerate radiative recombination and enhance luminescence performance via plasmonic effects, while also evaluating the emission efficiency and stability of thin films and light-emitting diode (LED)

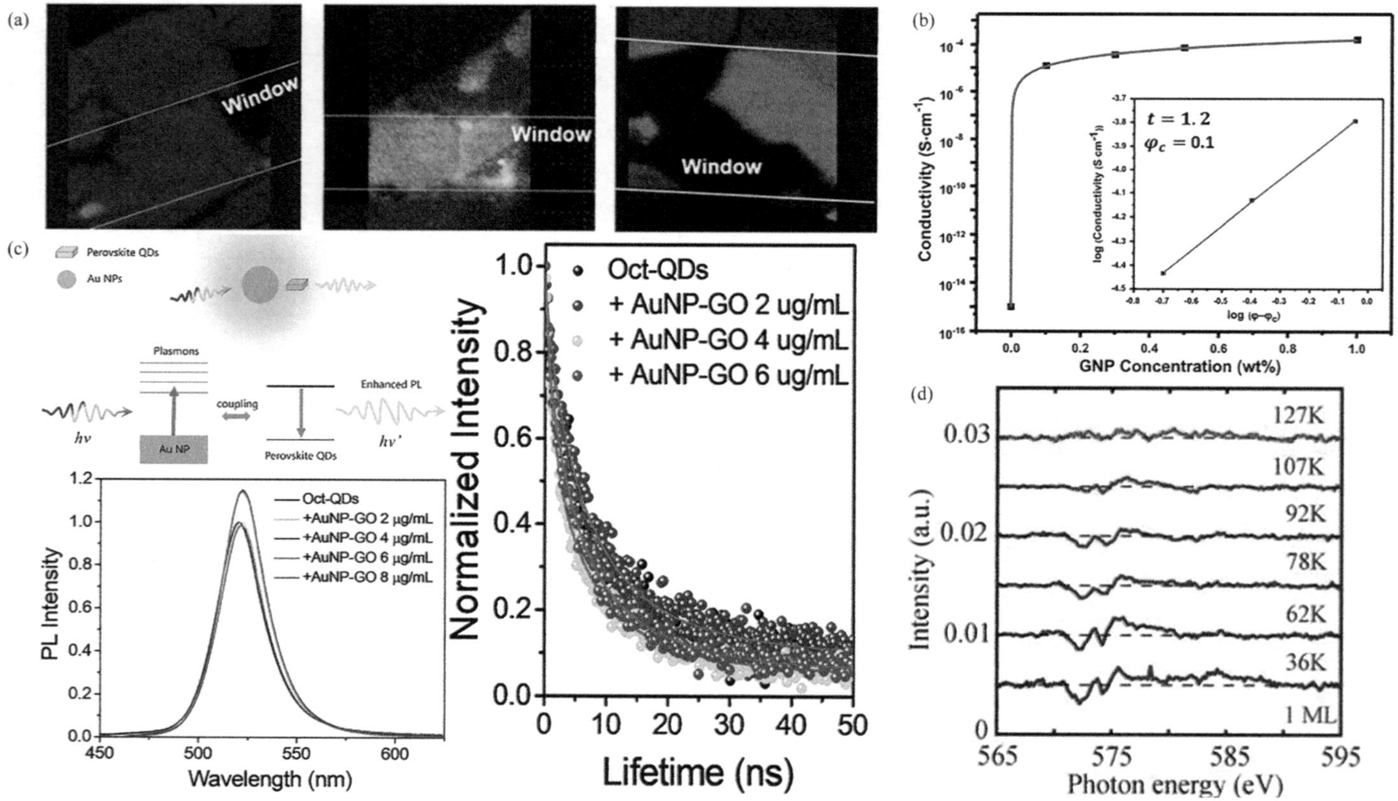

FIGURE 2.9 (a) PL mapping of the sample.[59] (b) Electrical characterization of the material using the four-probe method.[60] (c) PL spectra and PL decay of the sample.[61] (d) Ferromagnetic characterization of the material using X-ray magnetic circular dichroism (XMCD).[62]

devices.[61] The four-point probe measurement system is a specialized instrument for determining material resistivity and conductivity, operating on the four-point probe method principle. It applies a specific current to the sample, measures the corresponding voltage, and calculates electrical performance parameters through established formulas. This instrument features high measurement precision, relatively straightforward operation, and flexible requirements for sample geometry, and finds extensive applications in electrical characterization and research on semiconductor materials, conductive films, and various low-dimensional quantum conductive materials. Tejendra K et al. measured the bulk electrical conductivity of composites with varying GNP contents using the four-probe method, and analyzed the formation mechanism of conductive networks by calculating the percolation threshold and critical exponent t based on classical percolation theory.[60] Numerous exceptional properties of quantum materials originate from Coulomb correlations governing electronic quasiparticles and their interaction chains.

In low-dimensional quantum materials, performance enhancement primarily correlates with reduced dimensionality, thereby inducing novel phases such as ferroelectricity, ferromagnetism, and multiferroicity. XMCD is a powerful characterization technique that uses synchrotron radiation and left- and right-circularly polarized X-rays to selectively excite inner-shell electrons of specific elements. Due to spin–orbit coupling and the Pauli exclusion principle, X-ray absorption intensity depends on the relative orientation between X-ray polarization and sample magnetization; measuring the differential absorption of two polarizations enables quantitative determination of spin and orbital magnetic moments for specific elements. Compared with conventional magnetic characterization methods, this technique offers distinct advantages in low-dimensional quantum material research and provides critical mechanistic insights into spintronic devices such as magnetic tunnel junctions in spin valves. The elemental selectivity and high sensitivity of XMCD render it a potent tool for investigating multicomponent magnetic materials and magnetic interfaces. Yong Zhong et al. employed XMCD to detect the Cr $L_{2,3}$-edge magnetic circular dichroism signals in Cr_2Te_3 films of varying thicknesses, quantitatively determining the Curie temperature and relative magnetization. Their results confirm the existence of ferromagnetism in the 2D limit, reveal its thickness dependence, and corroborate the transition of the ferromagnetic ground state—providing critical magnetic evidence for the core conclusions.[62]

2.3 CHALLENGES AND PROSPECTS

The growth of high-quality low-dimensional quantum materials must satisfy four critical requirements: (i) precise and uniform control of dimensions and thickness to ensure consistent quantum confinement effects; (ii) minimized defect densities and surface contamination to reduce scattering of electrons/spins by defects and interference of quantum states by substrates, ligands, and impurities, thereby preserving quantum transport properties; (iii) ordered crystalline phases and structures, such as 3R/2H phases in 2D MX_2 and chirality in carbon nanotubes, which directly determine the stability of quantum properties including ferroelectricity and valleytronic characteristics; (iv) tunable experimental parameters with reproducible conditions, enabling precise control of crystalline phase, composition, and strain through process optimization to meet large-scale industrial production demands. The growth and structural control of low-dimensional quantum materials essentially constitute an optimization process within the parameter space, yet this adjustment faces significant challenges due to three primary constraints: incomplete information with unclear nucleation and growth mechanisms; non-comprehensive information given the vast parameter space precluding exhaustive exploration; and imperfect information arising from inevitable experimental errors. Under these constraints, machine learning demonstrates unique advantages by synthesizing massive experimental and computational data from historical literature and databases, extracting valuable insights to establish parameter-performance correlations, and achieving property prediction and optimization of experimental conditions through deep learning.

Under the growth, processing, and operational conditions of low-dimensional quantum materials, precise knowledge of specific reaction processes, underlying mechanisms, and the evolution of dynamic performance remains challenging. Structural control of low-dimensional materials depends on a thorough understanding of growth mechanisms, which in turn requires real-time observation of structure formation and evolution during growth. In situ techniques integrate structural characterization with performance measurement within a single instrument, enabling real-time observation of nucleation processes during material growth and tracking structural evolution under operational conditions, thereby directly establishing process-structure-performance relationships.

Different characterization techniques provide structural information across distinct material dimensionality. SEM probes surface structures, TEM reveals internal structural information, and STM captures minute topographic variations on flat surfaces. Diverse processes arising from identical interactions generate distinct structural information. Secondary or scattered electrons from electron beam scattering and interactions form SEM images, transmitted electrons produce TEM images, characteristic X-rays yield EDS, diffracted X-rays generate XRD patterns, and photoemitted electrons constitute XPS. A comprehensive understanding of low-dimensional quantum material structures, therefore, requires the integrated application of multiple characterization methods to avoid partial, "blind men and elephant" perspectives.

2.4 SUMMARY

This chapter presents a systematic discussion on the preparation and characterization of low-dimensional quantum materials. Leveraging quantum confinement effects, these materials exhibit unique physical properties. Their preparation encompasses three primary technical routes: vapor-phase, liquid-phase, and solid-phase synthesis. Vapor-phase deposition enables precise growth of multi-dimensional materials; liquid-phase synthesis offers advantages of low cost, high purity, and environmental compatibility; solid-phase methods include both top-down mechanical routes and solid-state reaction routes for producing thin layers and nanostructures. Characterization techniques have evolved into a comprehensive, multi-scale and multi-dimensional framework spanning microscopic morphology, crystal structure, chemical composition, and optoelectromagnetic properties, enabling thorough elucidation of structure-property relationships. Current research on low-dimensional quantum materials still faces critical challenges, including dimensional uniformity, defect control, crystalline phase regulation, and scalable reproducible fabrication. Future directions may involve machine learning for preparation parameter optimization, in situ characterization for real-time observation of growth kinetics, and multi-technique integration for comprehensive characterization, thereby advancing controlled synthesis of high-quality low-dimensional quantum materials and their practical applications in quantum sensing, spintronics, and related fields.

REFERENCES

[1] F. J. G. Silva, A. Baptista, J. Porteiro, J. L. Míguez, G. Pinto, L. Fernandes, On the physical vapour deposition (PVD): Evolution of magnetron sputtering processes for industrial applications. *Procedia Manufacturing*, 17, 746–757 (2018).

[2] Xizheng Wang, et al., Electrified vapour deposition at ultrahigh temperature and atmospheric pressure for nanomaterials synthesis. *Nature Synthesis*, 5, 14–26 (2026).

[3] Jacob Elkins, et al., Spatially controlled growth of ultrathin MoO2 polymorphs by physical vapor deposition. *Nano Letters*, 25, 2283–2289 (2025).

[4] Stefan Meister, Hailin Peng, Candace K. Chan, Xiao Feng Zhang, Yi Cui, Morphology control of layer-structured gallium selenide nanowires. *Nano Letters*, 7, 199–203 (2006).

[5] Haoxin Huang, et al., In-sensor compressing via programmable optoelectronic sensors based on van der Waals heterostructures for intelligent machine vision. *Nature Communications*, 16, 3836 (2025).

[6] Shahid Alam, Muhammad Zahir Iqbal, Nayyab Amjad, Rashid Ali, Meshal Alzaid, Magnetron sputtered tungsten di-sulfide: An efficient battery grade electrode for supercapattery devices. *Journal of Energy Storage*, 46, 103861 (2022).

[7] Emese Huszar, et al., Morphology and yield of W nanoparticles in gas aggregation: Pathways towards functional nanocomposite films. *Surface and Coatings Technology*, 521, 133089 (2026).

[8] Kang Li, et al., Ultrafast, groove-confined synthesis of ultrafine Pt nanocubes for efficient electrocatalytic ammonia oxidation. *Matter*, 9, 102566 (2026).

[9] Louise R. Smith, et al., Direct formation of the atomic Pd-ZnO interface by magnetron sputtering primed for methanol production from CO_2. *ACS Catalysis*, 15, 15502–15508 (2025).

[10] Yuhui Chen, Congbao Guo, Yi Wang, Kun Wang, Shuqin Song, Constructing high-entropy spinel oxide thin films via magnetron sputtering for efficiently electrocatalyzing alkaline oxygen evolution reaction. *Chinese Journal of Catalysis*, 77, 210–219 (2025).

[11] Jiashuai Yuan, et al., Controllable synthesis of nonlayered high-κ Mn_3O_4 single-crystal thin films for 2D electronics, *Nature Communications*, 16, 964 (2025).

[12] Chung-Jen Chen, et al., Tunable electron correlation in epitaxial 1T-TaS_2 spirals, *Advanced Materials*, 37, 2413926 (2025).

[13] R. S. Wagner, W. C. Ellis, Vapor-liquid-solid mechanism of single crystal growth. *Applied Physics Letters*, 4, 89–90 (1964).

[14] Stig Helveg, et al., Atomic-scale imaging of carbon nanofibre growth. *Nature*, 427, 426–429 (2004). DOI:10.1038/nature02278

[15] Mali Zhao, et al., Coherent thermoelectric power from graphene quantum dots. *Nano Letters*, 19, 61–68 (2018).

[16] Xiaoyue Wang, et al., Tunable photodetection range of GaSb nanowires for infrared spectral encryption. *Nano Letters*, 25, 15000–15006 (2025).

[17] Mingzhi Zou, et al., Electric-field-assisted PECVD for growth of high-purity horizontally aligned semiconducting carbon nanotube arrays. *Advanced Functional Materials*, 35, 2504150 (2025).
[18] Weiwei Cai, Xuesong Li, Jinho An, Seyoung Kim, Junghyo Nah, Dongxing Yang, Richard Piner, Aruna Velamakanni, Inhwa Jung, Emanuel Tutuc, Sanjay K. Banerjee, Luigi Colombo, Rodney S. Ruoff, Large-area synthesis of high-quality and uniform graphene films on copper foils. *Science*, 324, 1312–1314 (2009).
[19] F. Gonca Aras, et al., A review on recent advances of chemical vapor deposition technique for monolayer transition metal dichalcogenides (MX_2: Mo, W; S, Se, Te). *Materials Science in Semiconductor Processing*, 148, 106829 (2022).
[20] Jiashuo Yan, et al., CVD controlled preparation and growth mechanism of 2H-WS_2 nanosheets. *Vacuum*, 207, 111564 (2023).
[21] Kibum Kang, et al., High-mobility three-atom-thick semiconducting films with wafer-scale homogeneity. *Nature*, 520, 656–660 (2015). DOI:10.1038/nature14417
[22] Qianqian He, et al., Direct synthesis of controllable ultrathin heteroatoms-intercalated 2D layered materials. *Nature Communications*, 15, 6320 (2024).
[23] Satoru Fukamachi, et al., Large-area synthesis and transfer of multilayer hexagonal boron nitride for enhanced graphene device arrays. *Nature Electronics*, 6, 126–136 (2023).
[24] Zhiyuan Shi, et al., Vapor–liquid–solid growth of large-area multilayer hexagonal boron nitride on dielectric substrates. *Nature Communications*, 11, 849 (2020).
[25] Xuesong Li, et al., Graphene films with large domain size by a two-step chemical vapor deposition process. *Nano Letters*, 10, 4328–4334 (2010).
[26] Shisheng Li, Salt-assisted chemical vapor deposition of two-dimensional transition metal dichalcogenides. *iScience*, 24, 103229 (2021).
[27] Hao Huang, et al., Water-soluble monolayer molybdenum disulfide quantum dots with upconversion fluorescence. *Particle & Particle Systems Characterization*, 32, 72–79 (2014).
[28] Shixiu Cao, et al., Hydrothermal synthesis of variety low dimensional WS2 nanostructures, *Materials Letters*, 129, 205–208 (2014).
[29] Tasnimul Quader Tazim, Md Kawsar, Md Sahadat Hossain, Newaz Mohammed Bahadur, Samina Ahmed, Hydrothermal synthesis of nano-metal oxides for structural modification: A review. *Next Nanotechnology*, 7, 103229 (2025).
[30] Mariana C. F. Costa, Sergio G. Echeverrigaray, Daria V. Andreeva, Kostya S. Novoselov, Antonio H. Castro Neto, Two-dimensional quantum dots: From photoluminescence to biomedical applications. *Solids*, 3, 578–602 (2022).
[31] Chunxia Wang, et al., Dual functional hydrogen peroxide boosted one step solvothermal Synthesis of highly uniform sulfur quantum dots at elevated temperature and their fluorescent sensing. *Sensors and Actuators B: Chemical*, 344, 130326 (2021).
[32] Marwa Ben Arbia, Hicham Helal, Elisabetta Comini, Recent advances in low-dimensional metal oxides via sol-gel method for gas detection. *Nanomaterials*, 14, 359 (2024).
[33] Tal Zada, Meital Reches, Daniel Mandler, Antifouling and antimicrobial coatings based on sol–gel films. *Journal of Sol-Gel Science and Technology*, 95, 609–619 (2020).
[34] Ehsan Rahmani, Preparation and characterization of thin films by sol-gel method, thin films-growth. *Characterization and Electrochemical Applications*, 17, 1–13 (2024).
[35] Hui Yang, Mankang Zhu, Yue Li, Sol–gel research in China: A brief history and recent research trends in synthesis of sol–gel derived materials and their applications. *Journal of Sol-Gel Science and Technology*, 106, 406–421 (2022).

[36] Andreea Mariana Negrescu, et al., Metal oxide nanoparticles: Review of synthesis, characterization and biological effects. *Journal of Functional Biomaterials*, 13, 274 (2022).
[37] Moniruzzaman Syed, et al., Structural and electronic impact on various substrates of TiO_2 thin film using sol-gel spin coating method. *Journal of Materials Science Research*, 13, 1–14 (2024).
[38] Tian Hong Chen, Meng Di Hu, Hong Liang Zhu, Hao Liu, Yi Shi, and Shan Cheng Yan, Thin indium-tin-oxide film produced through sol-gel method with its morpholgical and electrical properties. *Science of Advanced Materials*, 14, 1298–1303 (2022).
[39] A. K. Geim, K. S. Novoselov, S. V. Morozov, D. Jiang, Y. Zhang, S. V. Dubonos, I. V. Grigorieva, A. A. Firsov, Electric field effect in atomically thin carbon films. *Science*, 306, 666–669 (2004).
[40] Xin Lu, Synthesis and optical properties of atomically thin transition metal dichalcogenides. *Nanophotonics*, 6, 1289–1308 (2017).
[41] Małgorzata Giza, et al., Exploring the application of gold-assisted exfoliation in large-scale integration of n-type and p-type 2D-FETs. *Small Methods*, 10, 2500559 (2026). DOI: 10.1002/smtd.202500559
[42] Yuan Huang, et al., Reliable exfoliation of large-area high-quality flakes of graphene and other two-dimensional materials. *ACS Nano*, 9, 10612–10620 (2015).
[43] Matej Velicky, et al., Mechanism of gold-assisted exfoliation of centimeter-sized transition-metal dichalcogenide monolayers. *ACS Nano*, 12, 10463–10472 (2018). DOI:10.1021/acsnano.8b06101
[44] Yunrui Yan, Exfoliation of hexagonal boron nitride nanosheets assisted with covalent organic frameworks by ball-milling. *Journal of Physics: Conference Series*, 2338, 012055 (2022).
[45] M. Sherif El-Eskandarany, et al., Mechanical milling: A superior nanotechnological tool for fabrication of nanocrystalline and nanocomposite materials. *Nanomaterials*, 11, 2484 (2021).
[46] H. L. Shi, et al., Structural properties and superconductivity of $SrFe_2As_{2-x}P_x (0.0 \leq x \leq 1.0)$ and $CaFe_2As_{2-y}P_y (0.0 \leq y \leq 0.3)$. *Journal of Physics: Condensed Matter*, 22, 125702 (2010).
[47] Gul Amin, *ZnO and CuO Nanostructures: Low Temperature Growth, Characterization, Their Optoelectronic and Sensing Applications*. Linkopings Universitet, Sweden (2012).
[48] Naresh Marturi, Vision and visual servoing for nanomanipulation and nanocharacterization in scanning electron microscope. PhD Diss., Université de Franche-Comté (2013).
[49] Wonhee Ko, Chuanxu Ma, Giang D. Nguyen, Marek Kolmer, An-Ping Li, Atomic-scale manipulation and in situ characterization with scanning tunneling microscopy. *Advanced Functional Materials*, 29, 1903770 (2019).
[50] Ying Ying Wang, et al., Thickness identification of two-dimensional materials by optical imaging. *Nanotechnology*, 23, 495713 (2012).
[51] Qingge Wang, et al., Microstructure and mechanical properties of SiC whiskers modified Ti-39Nb alloy fabricated by laser powder bed fusion. *Materials Science and Engineering: A*, 922, 147608 (2025).
[52] Sarah K. Adams, Nicholas W. Piekiel, Matthew H. Ervin, Christopher J. Morris, Silicon quantum dots for energetic material applications. *Applied Physics Letters*, 112, 233108 (2018).
[53] Wen-Peng Han, et al., The numerical-aperture-dependent optical contrast and thickness determination of ultrathin flakes of two-dimensional atomic crystals: A case of graphene multilayers. *Acta Physica Sinica*, 62, 110702 (2013).
[54] Qishuo Yang, et al., Unified transmission electron microscopy with the glovebox integrated system for investigating air-sensitive two-dimensional quantum materials. *The Innovation*, 6, 100751 (2025).

[55] Yaqian Wang, et al., Spin-valley locking effect in defect states of monolayer MoS_2. *Nano Letters*, 20, 2129–2136 (2020).
[56] Shengdi Zhao, et al., Elucidating atomic structure and reconstruction of Mn_3O_4(001) surface. *Applied Surface Science*, 680, 161339 (2025).
[57] Maryam Abdolahpour Salari, Volkan Şenay, Günay Merhan Muğlu, Sevda Sarıtaş, Mutlu Kundakçı, Evaluation of the optical, structural, and morphological characteristics of a Sn-doped α-Fe_2O_3 thin film fabricated using RF and DC magnetron Co-sputtering technique. *Ceramics International*, 51, 23068–23076 (2025).
[58] Dong Hyun Kim, et al., In-situ thermal phase transition and structural investigation of ferroelectric tetragonal barium titanate nanopowders with pseudo-cubic phase. *Chemosphere*, 283, 131218 (2021).
[59] Nengjie Huo, et al., Improved photoluminescence in 2D semiconductors induced by interface magnetization. *ACS Photonics*, 7, 3341–3345 (2020).
[60] Tejendra K. Gupta, M. Choosri, K. M. Varadarajan, S. Kumar, Self-sensing and mechanical performance of CNT/GNP/UHMWPE biocompatible nanocomposites. *Journal of Materials Science*, 53, 7939–7952 (2018).
[61] Gautham Kumar, Chien-Chung Lin, Hao-Chung Kuo, Fang-Chung Chen, Enhancing photoluminescence performance of perovskite quantum dots with plasmonic nanoparticles: Insights into mechanisms and light-emitting applications. *Nanoscale Advances*, 6, 782–791 (2024).
[62] Yong Zhong et al., From stoner to local moment magnetism in atomically thin Cr_2Te_3. *Nature Communications*, 14, 5340 (2023).

3 Basic Physical Properties of Low-Dimensional Quantum Materials

Lihong Zhou and Zongyu Huang

3.1 TIGHT-BINDING MODEL

In solid-state physics, understanding how the discrete energy levels of isolated atoms evolve into the continuous energy bands of a crystal is fundamental to explaining macroscopic material properties such as electrical conductivity and optical response. Band theory provides this framework, and the tight-binding (TB) model is one of the most intuitive approximate methods for constructing band structures. Its core physical picture is that electrons are primarily bound to their respective atomic nuclei but have a finite probability of hopping to the orbitals of neighboring atoms via quantum tunneling. This interatomic interaction causes the broadening and shifting of energy levels, ultimately forming energy bands.

Owing to its clear conceptual foundation and relatively simple mathematical form, the TB model serves as an essential tool in both teaching and research. It is particularly well-suited for materials with strongly localized electrons, such as transition–metal oxides, organic crystals, and the increasingly prominent class of two-dimensional (2D) materials (e.g., graphene and transition metal dichalcogenides). This chapter aims to systematically expound the fundamental principles of the TB model and demonstrate its concrete application and computational procedure through the paradigmatic case of graphene. This chapter first introduces the basic theory of the TB model, including the derivation of its Hamiltonian, the application of Bloch's theorem, and general methods for band calculation.

3.1.1 Lattice Model

Solid materials consist of a vast number of atoms (or ions), with approximately 10^{23} atoms per cubic centimeter. This enormous number of atoms is arranged in a specific manner, and the way the atoms are organized is referred to as the structure of the solid. Traditionally, solid materials have been classified into three broad categories: crystalline, non-crystalline (amorphous), and quasi-crystalline. In an ideal crystal, the atomic arrangement is highly regular and periodic. The atomic arrangement in solids serves as the foundation for studying the macroscopic properties and various microscopic processes of solid materials. The specific arrangement of atoms in a

DOI: 10.1201/9781003655305-4

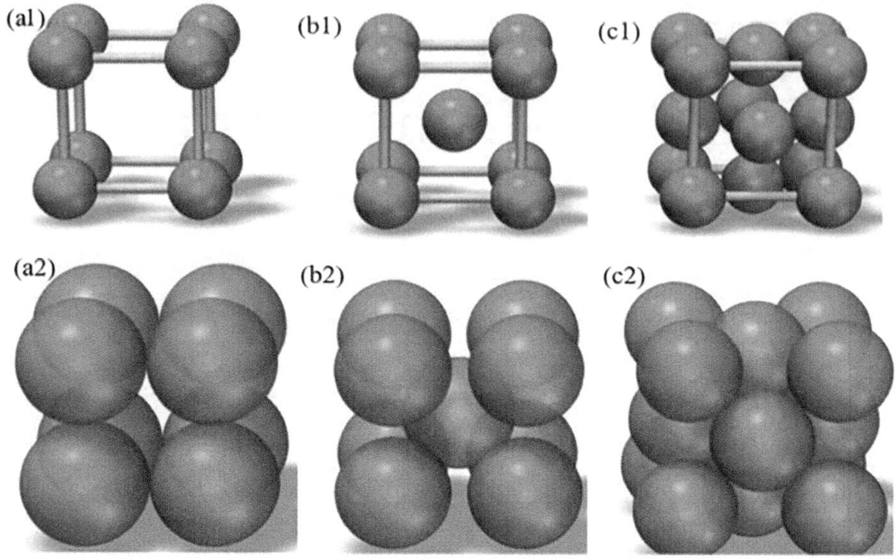

FIGURE 3.1 Atomic arrangement. (a1) and (a2) The primitive cubic (simple cubic [SC]). (b1) and (b2) Body-centered cubic (BCC). (c1) and (c2) Face-centered cubic (FCC).

crystal is generally referred to as the crystal lattice. Different crystals may have distinct atomic arrangements. Figure 3.1 depicts three common cubic lattices: the SC, BCC, and FCC.

In nature, almost no actual crystals possess an ideal SC lattice structure, where atoms are positioned solely at the corners of the cube, like geometric points (as shown in Figure 3.1a). However, more complex lattice structures can be analyzed and understood as extensions of the SC lattice. For instance, a BCC lattice can be constructed by adding an atom at the body center of a SC lattice. A FCC lattice is formed by adding an atom at the center of each cube face of a SC lattice. Many metals, such as Li, Na, K, Rb, Cs, and Fe, possess a BCC structure, while Au, Ag, Cu, and Al exhibit a FCC lattice. The diamond lattice, formed by carbon atoms, represents another fundamental crystal structure. Its conventional unit cell is illustrated in Figure 3.2a. It can be described by taking a FCC unit cell, drawing eight diagonals from its center to the corners, and placing an atom at the midpoint of four non-adjacent diagonals. A key structural feature is that each atom has four nearest neighbors, which are located precisely at the corners of a regular tetrahedron.

Several common compound crystal structures are fundamental in materials science. The GaSn structure is analogous to the diamond lattice, but with two different atom types: one occupies the FCC positions, while the other occupies half of the tetrahedral sites (e.g., along body diagonals), as shown in Figure 3.2b. Many III–V semiconductors, such as InAs and InSb, crystallize in this structure. The NaCl structure can be described as two interpenetrating FCC lattices, one for cations (Na^+) and one for anions (Cl^-), offset by half a cube edge along the axes as shown in Figure 3.2c. This results in each ion being octahedrally coordinated by six ions of the opposite

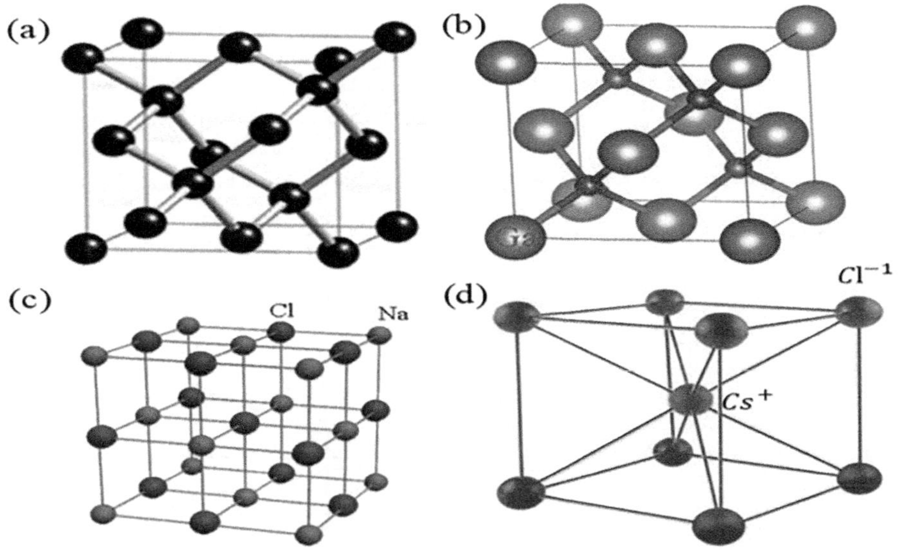

FIGURE 3.2 (a) The structure of diamond, (b) zinc blende (GaAs) structure, (c) rock salt (NaCl) structure, and (d) cesium chloride (CsCl) structure.

charge. With the exception of CsCl, CsBr, and CsI, compounds composed of alkali metals (Li, Na, K, and Rb) and halogens (F, Cl, Br, and I) typically adopt the NaCl crystal structure. The CsCl structure is a variant of the BCC lattice (Figure 3.2d), where one type of atom (Cs^+) sits at the cube corners, and the other (Cl^-) resides at the cube center, giving each ion eight nearest neighbors of the opposite type. The lattice types mentioned are standard and widely studied structures in solid-state physics.

All crystal lattices possess periodicity, which is typically described using a primitive cell and basis vectors in Figure 3.3a. Geometrically, a lattice is used to represent the periodicity of the crystal structure, and the position of any lattice point is given by a lattice vector:

$$\boldsymbol{R}_l = l_1\boldsymbol{a}_1 + l_2\boldsymbol{a}_2 + l_3\boldsymbol{a}_3 \tag{3.1}$$

where $\boldsymbol{a}_1, \boldsymbol{a}_2,$ and $\boldsymbol{a}_3$ are the primitive translation vectors (basis vectors) and $l_1, l_2,$ and l_3 are integers.

The resulting parallelepiped is known as the elementary cell or primitive cell. In three dimensions (3D), the volume of the primitive cell is given by the scalar triple product of the basis vectors:

$$\Omega = \boldsymbol{a}_1 \cdot (\boldsymbol{a}_2 \times \boldsymbol{a}_3) \tag{3.2}$$

In solid-state physics, another commonly used primitive cell with a more symmetric shape is the W–S cell (as shown in Figure 3.3b). It is constructed as the region of

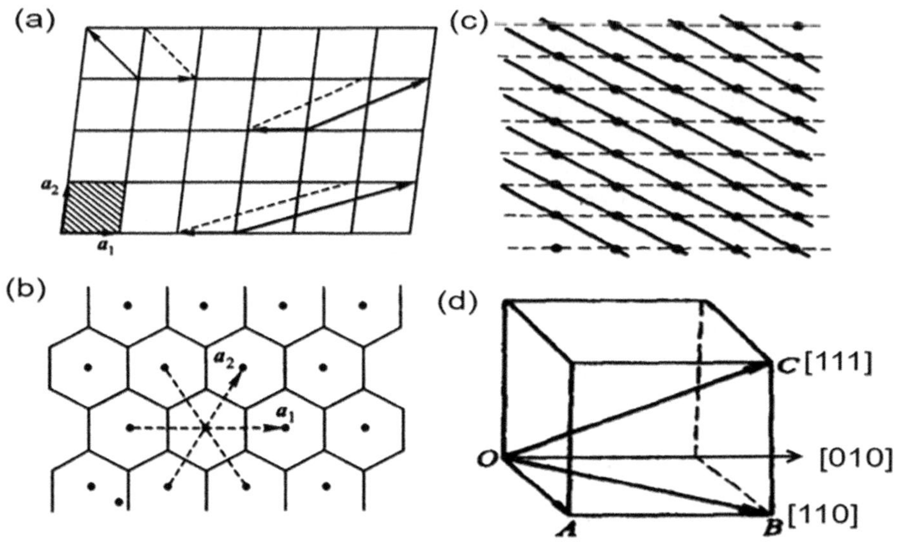

FIGURE 3.3 (a) The primitive cell and basis vectors in a two-dimensional (2D) lattice. (b) Wigner–Seitz (W–S) cell of a hexagonal lattice.[1] (c) Crystal row. (d) Crystallographic direction of the SC.[2]

space that is closer to a given lattice point than to any other. This is done by drawing the perpendicular bisector planes of the lines connecting the point to its nearest (and sometimes next-nearest) neighbors. The resulting polyhedron contains exactly one lattice point. The W–S cell of the SC lattice is a cube, while that of the BCC lattice is a truncated octahedron. In contrast, the W–S cell of the FCC lattice is a rhombic dodecahedron. A key property of the W–S cell is its symmetry: it exhibits the full symmetry of the crystal's point group. This means that all symmetry operations (rotations, reflections, and inversions) of the lattice also map the W–S cell onto itself. Consequently, it serves as an ideal and convenient choice for a primitive cell in theoretical analyses.

For theoretical convenience, a second, artificially constructed lattice is introduced in wave-vector space, known as the reciprocal lattice. Its fundamental vectors have the dimension of inverse length relative to those in real coordinate space. The primitive vectors of the reciprocal lattice are defined as follows:

$$\boldsymbol{a}_i \cdot \boldsymbol{b}_j = 2\pi\delta_{ij}, \left(i, j = 1,2,3\right) \tag{3.3}$$

Then we can get the reciprocal lattice basis:

$$\boldsymbol{b}_i = \frac{2\pi}{\Omega} \epsilon_{ijk} \left(\boldsymbol{a}_j \times \boldsymbol{a}_k\right) \tag{3.4}$$

The reciprocal lattice vector is written as

$$\boldsymbol{K}_n = n_1\boldsymbol{b}_1 + n_2\boldsymbol{b}_2 + n_3\boldsymbol{b}_3 \tag{3.5}$$

The key relationship between the direct and reciprocal lattice vectors is written as

$$\boldsymbol{K}_n \cdot \boldsymbol{R}_l = 2\pi \sum_{i=1}^{3} n_i l_i = 2\pi m, \; (m = \text{integer}) \tag{3.6}$$

To analyze the effect of crystal symmetry on wave-like excitations (e.g., electrons and phonons), it is standard to construct the W–S cell of the reciprocal lattice. This special W–S cell, when centered on a reciprocal lattice point (conventionally taken as the origin), defines the first Brillouin zone (BZ). The BZ is a symmetric primitive cell in reciprocal space, which retains the complete point-group symmetry of the direct lattice.

Another fundamental characteristic of crystals is their anisotropy—their properties vary along different lattice directions. The lattice points of a Bravais lattice can be viewed as distributed across a set of parallel lines, known as crystal rows, as illustrated in Figure 3.3c. The solid and dashed lines represent two distinct crystal rows. The same lattice can be partitioned into rows along different orientations, each defining a specific direction called the crystallographic direction. This direction can be represented by integer indices derived from the lattice vector l_1, l_2, and l_3, which is represented as $\left[l_1 l_2 l_3\right]$ (Figure 3.3d). Owing to anisotropy, a crystal's physical properties (including conductivity, mechanical behavior, and chemical reactivity) vary markedly with crystallographic direction. Hence, the identification and notation of these directions form a basis for structural, functional, and morphological analysis.

Amorphous materials, in contrast to their crystalline counterparts, lack long-range periodic order in their atomic structure. Instead, they exhibit only short-range order. The radial distribution function, measured via diffraction techniques (X-ray, electron, or neutron), is the primary experimental method for characterizing this short-range order and analyzing the structure of amorphous materials.

Beyond the crystalline and amorphous states, solids can form a quasi-crystalline phase. Quasi-crystals possess long-range orientational order, with point-group symmetries forbidden in periodic crystals, but lack long-range translational symmetry. Along symmetry axes, they exhibit quasi-periodicity, resulting from a specific arrangement of two or more incommensurate characteristic length scales.

3.1.2 Bloch's Theorem and Band Structure Calculation

The periodicity of a crystal lattice refers to the property that its atoms or structural units are arranged in space in a specific, regular, and repeating manner. This periodicity is a manifestation of translational symmetry and forms the fundamental basis for understanding most of the physical properties of crystals.

Under the Born-von Karman boundary conditions, a crystal composed of N primitive cells exhibits discrete translational symmetry. Correspondingly, the wave vector k in reciprocal space becomes quantized, and its allowed values, determined by the boundary conditions, typically take the following form:

$$\boldsymbol{k} = \sum_{j=1}^{3} \frac{n_j}{N_j} \boldsymbol{b}_j \tag{3.7}$$

For any translation by a direct lattice vector $\boldsymbol{R}_1$, we have

$$\{E \mid \boldsymbol{R}_1\}^{-1} \psi(\boldsymbol{r}) = \psi(\boldsymbol{r} + \boldsymbol{R}_1) = e^{ik \cdot \boldsymbol{R}_1} \psi(\boldsymbol{r}) \tag{3.8}$$

The preceding formula is interpreted as the eigenvalue equation for the translation operator. The exponential factor $e^{ik \cdot \boldsymbol{R}_1}$ represents the kth eigenvalue, and its associated eigenfunction is denoted accordingly by $\psi_k(\boldsymbol{r})$. Therefore, the expression can be written in the form

$$\{E \mid \boldsymbol{R}_1\}^{-1} \psi(\boldsymbol{r}) = \psi(\boldsymbol{r} + \boldsymbol{R}_1) = e^{ik \cdot \boldsymbol{R}_1} \psi_k(\boldsymbol{r}) \tag{3.9}$$

which is Bloch's theorem; this states that in a potential with lattice periodicity, the single-electron wavefunction can be written as the product of a plane wave and a function with the same periodicity (known as a Bloch function),

$$\psi_{nk}(\boldsymbol{r}) = e^{ik \cdot \boldsymbol{r}} u_{nk}(\boldsymbol{r}) \tag{3.10}$$

where n is the band index, and the amplitude function satisfies the relation $u_{nk}(\boldsymbol{r} + \boldsymbol{R}_l) = u_{nk}(\boldsymbol{r})$.

According to the Bloch theorem and lattice period, we can get the energy band. The Schrodinger equation can be written as

$$i\hbar \frac{\partial \Psi}{\partial t} = (H_0 + U)\Psi \tag{3.11}$$

where the single-particle Hamiltonian H_0 is

$$H_0 = -\frac{\hbar^2}{2m} \nabla^2 + V(\boldsymbol{r}) \tag{3.12}$$

with a periodic lattice potential $V(\boldsymbol{r} + 1) = \boldsymbol{V}(\boldsymbol{r})$. To address the problem above, the wavefunction Ψ can be expanded using the complete set of eigenfunctions determined by the ideal crystal Hamiltonian H_0. An alternative complete set for describing localized states is the set of Wannier functions, which are defined in terms of Bloch functions. The expansion of Ψ in terms of Wannier functions constitutes the Wannier representation.

Owing to its periodicity in the reciprocal lattice, a Bloch function $\Psi_{nk}(\boldsymbol{r})$ can be expanded into a Fourier series using the direct lattice vectors as

$$\Psi_{nk}(\boldsymbol{r}) = \frac{1}{\sqrt{N}} \sum_{\boldsymbol{l}} e^{ik \cdot \boldsymbol{l}} \boldsymbol{a}_n(\boldsymbol{r}, \boldsymbol{l}) \tag{3.13}$$

with

$$\frac{1}{\sqrt{N}} \sum_{k \in BZ} e^{ik \cdot (l - l')} = \delta_{ll'} \tag{3.14}$$

The inverse transformation is

$$a_n(\boldsymbol{r}, \boldsymbol{l}) = \frac{1}{\sqrt{N}} \sum_{k \in BZ} e^{\ ik\ l} \Psi_{nk}(\boldsymbol{r}) \tag{3.15}$$

According to Bloch's theorem, we can get

$$a_n(\boldsymbol{r}, \boldsymbol{l}) = a_n(\boldsymbol{r} - \boldsymbol{l}) \tag{3.16}$$

Wannier functions are orthogonal

$$\int a_n^{\star}(\boldsymbol{r} - \boldsymbol{l}) a_{n'}(\boldsymbol{r} - \boldsymbol{l}') = \delta_{nn'} \delta_{ll'} \tag{3.17}$$

and are complete

$$\sum_n \sum_l a_n^{\star}(\boldsymbol{r} - l) a_n(\boldsymbol{r}' - l) = \delta(\boldsymbol{r} - \boldsymbol{r}') \tag{3.18}$$

Therefore, Wannier functions can be used as basic functions to represent the wavefunction $\Psi(\boldsymbol{r})$ as

$$\Psi(\boldsymbol{r}) = \sum_{n,l} C_{nl} \boldsymbol{a}_n(\boldsymbol{r} - \boldsymbol{l}) \tag{3.19}$$

where C_{nl} is the annihilation operator in the Wannier representation, which annihilates an electron in the localized state of band n and site l. This formulation is also referred to as the local representation of the state vector. By taking Eq. (3.19) into the functions, we obtain the equation:

$$\sum_{l'} \left[\epsilon_n \left((\boldsymbol{l} - \boldsymbol{l}') \right) - E \delta_{l'} \right] C_{nl'} + \sum_{n', l'} U_{nn'}(\boldsymbol{l}, \boldsymbol{l}') C_{n'l'} = 0 \tag{3.20}$$

with

$$U_{nn'}(l,l') = \int a_n^*\big((r-1)\big)U(r)a_{n'}\big((r-1')\big)d(r) \tag{3.21}$$

If only a few matrix elements of U are non-zero, the equation can be solved directly. In the general case, however, the solution becomes computationally demanding, requiring extensive calculations on large atomic clusters. Although Wannier functions are not eigenstates of the Hamiltonian H, they are well suited to describe localized physical properties. We, therefore, adopt a TB approximation constructed from these Wannier functions.

$$\Psi_{nk}(r) = \sum_l C_{nk,l}a_n(r-1) \tag{3.22}$$

The equation of the coefficient $C_{nk,\alpha}$ is

$$\sum_\beta H_{\alpha\beta}(k)C_{nk,\beta} = E_{nk}C_{nk,\alpha} \tag{3.23}$$

where $H_{\alpha\beta}(k)$ is the matrix cell

$$H_{\alpha\beta}(k) = \epsilon_\alpha\delta_{\alpha\beta} + \sum_{R\in neighbors} t_{\alpha\beta}(R)e^{ik\cdot R} \tag{3.24}$$

Here, the summation runs over all neighboring atoms that have a hopping integral with the reference atom (the position vector difference is $\boldsymbol{R}$). By diagonalizing this k-dependent matrix $H(k)$, we obtain the dispersion relation E_{nk}, i.e., the band structure.

3.2 SUPERCONDUCTOR

The initial discovery of superconductivity dates to 1911, when Heike Kamerlingh Onnes and his colleagues at Leiden University observed that mercury lost all electrical resistance below 4.2 K, as shown in Figure 3.4. Superconductivity is a macroscopic quantum phenomenon in which a material transitions into a new state of matter below its critical temperature, characterized by the fundamental properties with zero electrical resistance. A material displaying this property is termed a superconductor. Superconductivity exhibits another fundamental property: the expulsion of magnetic flux (perfect diamagnetism).

Furthermore, experiments confirmed that the electronic specific heat of tin exhibits an exponential dependence on temperature in the superconducting state (see Figure 3.5a). The appearance of the exponential factor $e^{-\Delta/k_BT}$ in the specific heat curve indicates that single-particle excitations in the superconducting state require a minimum energy of Δ, demonstrating the characteristic energy gap of a superconductor. Moreover, since the superconducting phase has a lower free energy below T_c, a critical field H_c is necessary to suppress superconductivity and revert the metal to its normal, resistive state, illustrated in Figure 3.5b.

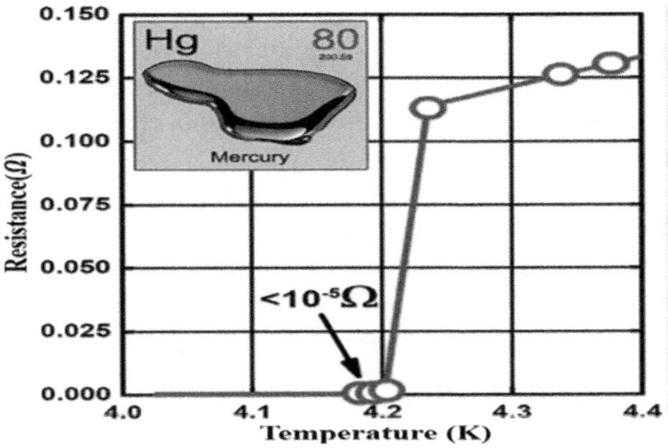

FIGURE 3.4 The resistance as a function of temperature for mercury Hg.[3]

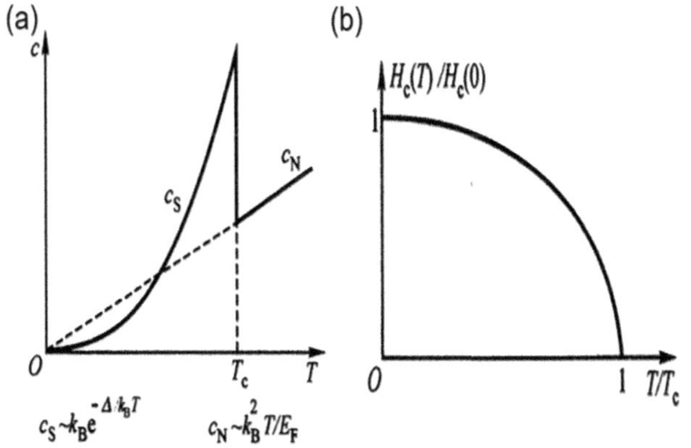

FIGURE 3.5 (a) Electronic specific heat capacity as a function of temperature. (b) Magnetic field as a function of temperature.

Based on their response to an external magnetic field, superconductors are classified into two types: Type I and Type II.[4] Among the discovered elemental metal superconductors, many belong to Type I, whereas elements such as vanadium, niobium, and technetium belong to Type II. Most alloy and compound superconductors are Type II, with some copper-oxide high-temperature superconductors classified as extreme (or hard) Type II superconductors.[5]

Based on their critical temperatures, superconductors are classified as low-temperature superconductors and high-temperature superconductors. To date, at ambient pressure, all superconductors that conform to the conventional Bardeen–Cooper–Schrieffer (BCS) theory have critical temperatures below 40 K. Only copper oxides and iron-based superconductors exceed 40 K at ambient pressure, and they are collectively referred to as "high-temperature superconductors." Under high pressure, however, the 40 K limit does not apply. For example, even BCS superconductors, such as certain metallic hydrides, can achieve critical temperatures above 200 K under high pressure. While Superconductors are categorized by material type as elemental superconductors aluminum alloy superconductors, intermetallic compound superconductors, oxide superconductors, organic superconductors, and heavy-fermion superconductors.

Research on superconductors is advancing significantly. On the one hand, a growing number of novel superconducting materials are being discovered. On the other hand, studies on their physical properties and underlying microscopic mechanisms have greatly propelled the development of condensed matter physics. Owing to their unique physical properties, superconductors hold important applications across multiple fields, including the energy and power sector, healthcare, transportation, fundamental scientific research, precision measurement, and quantum computing.

3.2.1 BCS Theory

The BCS theory, introduced in 1957 by John Bardeen, Leon Cooper, and Robert Schrieffer, explained superconductivity through the coherent condensation of Cooper pairs. According to the theory, electrons in the metal form pairs via an attractive interaction mediated by lattice vibrations. These pairs condense into a macroscopic quantum state, whose collective motion constitutes a lossless supercurrent. The theory successfully predicted the energy gap required to break these pairs, providing a comprehensive explanation for superconductivity.

The attractive interaction between electrons with opposite quasi-momenta and spins near the Fermi surface is the primary cause of superconducting condensation, which is described by the following model Hamiltonian:

$$H=\sum_{k}\epsilon_k\left(C_{k,\uparrow}^{\dagger}C_{k,\uparrow}+C_{-k,\downarrow}^{\dagger}C_{-k,\downarrow}\right)-V\sum_{k,k'}C_{k',\uparrow}^{\dagger}C_{-k',\downarrow}^{\dagger}C_{-k,\downarrow}C_{k,\uparrow} \tag{3.25}$$

with $\epsilon_k = E_k - \mu$, $E_k = \dfrac{\hbar^2 \boldsymbol{k}^2}{2m}$ is the single particle energy, μ is the chemical potential.

Since the terms $(\boldsymbol{k},\uparrow)$ and $(-\boldsymbol{k},\downarrow)$ in the expression always appear together, we can introduce a simplified notation by letting $\boldsymbol{k}$ represent $(\boldsymbol{k},\uparrow)$, $-\boldsymbol{k}$ represent $(-\boldsymbol{k},\downarrow)$. Then the Hamiltonian reduces to

$$H=\sum_{k}\epsilon_k\left(C_{k}^{\dagger}C_{k}+C_{-k}^{\dagger}C_{-k}\right)-V\sum_{k,k'}C_{k'}^{\dagger}C_{-k'}^{\dagger}C_{-k}C_{k} \tag{3.26}$$

According to the superposition principle, the eigenstates of the Hamiltonian H should consist of a linear combination of all possible scattering states,

$$|\psi\rangle=\sum_{k>k_F}a(\boldsymbol{k})C_{k}^{\dagger}C_{-k}^{\dagger}|F\rangle \tag{3.27}$$

with $|F\rangle$ represents the normal electronic ground state with a filled Fermi sphere, $a(\boldsymbol{k})$ is the coefficient. The energy is obtained by enforcing the normalization condition and carrying out the detailed calculation,[1]

$$E = -2\hbar\omega_D exp\left[-\frac{2}{g(0)V}\right] < 0 \tag{3.28}$$

The energy E corresponds to the ground-state energy of two electrons interacting via a weak attraction. A negative value ($E < 0$) signals the formation of a Cooper pair—a bound state of two electrons with opposite momenta and spins. Crucially, such binding occurs for any net attractive interaction between a pair of electrons outside the Fermi sphere, regardless of its strength. This implies that the energy of the bound pair lies below that of two free electrons at the Fermi level, rendering the filled Fermi sphere unstable in the presence of an attractive potential.

This section provides a brief introduction to solving the eigenfunctions and eigenvalues of a many-body Hamiltonian by employing mean-field theory. Assuming the ground-state expectation value of the pairing operator exists and is nonzero, it can be decomposed as follows:

$$C_{-k}C_k = \langle C_{-k}C_k\rangle + (C_{-k}C_k - < C_{-k}C_k >) \tag{3.29}$$

$$C_k^\dagger C_{-k}^\dagger = \langle C_k^\dagger C_{-k}^\dagger\rangle + (C_k^\dagger C_{-k}^\dagger - < C_k^\dagger C_{-k}^\dagger >) \tag{3.30}$$

Under the assumption that the deviation of the pair operator from its average is a small value, the interaction term is replaced by the expression above. Then the Hamiltonian of the self-consistent field approximation takes the form:

$$\begin{aligned}\bar{H} = \sum_k \epsilon_k\left(C_k^\dagger C_k + C_{-k}^\dagger C_{-k}\right) - V\sum_{k,k'}\{C_{k'}^\dagger C_{-k'}^\dagger\langle C_{-k}C_k\rangle \\ + C_{-k}C_k\langle C_{k'}^\dagger C_{-k'}^\dagger\rangle - \langle C_{k'}^\dagger C_{-k'}^\dagger\rangle\langle C_{-k}C_k\rangle\}\end{aligned} \tag{3.31}$$

the order parameter is defined as

$$\begin{aligned}\Delta &= V\sum_k \langle C_{-k}C_k\rangle \\ \Delta^* &= V\sum_k \langle C_k^\dagger C_{-k}^\dagger\rangle\end{aligned} \tag{3.32}$$

In general, Δ is a complex quantity. For an isolated, bulk sample system, we can take a gauge transformation as $\Delta = |\Delta| e^{i\phi}$ and $\Delta^* = |\Delta| e^{-i\phi}$, the phase ϕ can be an arbitrary constant. In order to simplify the calculation, take $\phi = 0$ we can get $\Delta = \Delta^* = |\Delta|$ as a real number. The Hamiltonian is

$$\bar{H} \approx \sum_k \epsilon_k \left(C_k^\dagger C_k + C_{-k}^\dagger C_{-k}\right) - \Delta \sum_k \left(C_k^\dagger C_{-k}^\dagger + C_{-k} C_k\right) + \frac{\Delta^2}{V} \tag{3.33}$$

The Hamiltonian can be diagonalized via the Bogoliubov canonical transformation,

$$\bar{H} = E_s(0) + \sum_k \sqrt{\epsilon_k^2 + \Delta^2}\left(\alpha_k^\dagger \alpha_k + \alpha_{-k}^\dagger \alpha_{-k}\right) \tag{3.34}$$

with the ground-state energy

$$E_s(0) = 2\sum_k \xi_k v_k^2 - 2\Delta \sum_k u_k v_k + \frac{\Delta^2}{V} \tag{3.35}$$

and $\xi_k = \sqrt{\epsilon_k^2 + \Delta^2}$ is the quasi-particle excitation energy. The existence of a finite energy gap at the Fermi level, which is the minimum quasi-particle excitation energy, is a key signature of superconductivity that is not found in normal metals.

The relations for the quasi-particle operators

$$\begin{gathered} \alpha_k = u_k C_k - v_k C_{-k}^\dagger, \alpha_k^\dagger = u_k C_k^\dagger - v_k C_{-k} \\ \alpha_{-k} = u_k C_{-k} + v_k C_k^\dagger, \alpha_{-k}^\dagger = u_k C_{-k}^\dagger + v_k C_k \end{gathered} \tag{3.36}$$

with the relations

$$\begin{aligned} u_k^2 &= \frac{1}{2}\left(1 + \frac{\epsilon_k}{\xi_k}\right) \\ v_k^2 &= \frac{1}{2}\left(1 - \frac{\epsilon_k}{\xi_k}\right) \end{aligned} \tag{3.37}$$

The BCS theory has been remarkably successful in explaining conventional superconductivity in most metals and alloys. However, it fails to account for the unconventional superconducting behavior observed in materials such as copper-based high-temperature superconductors and iron-based superconductors. In these systems, superconductivity likely originates from unconventional pairing mediated by electron correlation effects, such as spin fluctuations, rather than the traditional electron–phonon mechanism.

3.3 STRONG CORRELATED SYSTEM

While conventional band theory, built on the single-electron approximation, neglects electron–phonon coupling and treats valence electrons as independent particles, electron correlations become essential in narrow bands. The key reason is that an electron hopping between localized atomic orbitals must account for the target orbital's potential occupancy. If that orbital is already occupied, the strong on-site Coulomb repulsion between the two electrons must be incorporated. This incorporation leads to profound modifications of the electronic band structure.

Band theory frequently breaks down when describing transition–metal compounds, a classic example is MnO. While it predicts a metallic 3*d* band for MnO, the material is experimentally measured to be an insulator with very low conductivity. This fundamental failure stems from the theory's neglect of strong electron correlation effects in narrow bands.

3.3.1 Hubbard Model

The Hubbard model is the canonical minimal model for strongly correlated electrons, defined by only two terms: the kinetic energy of electron hopping and the on-site Coulomb repulsion. This fundamental competition between them drives a wealth of quantum phenomena, most notably the metal–insulator transition, and serves as the foundation for theoretical studies of high-temperature superconductivity and other forefront problems.

We consider a many-body system of a simple crystal with N atoms. Its Hamiltonian is written as follows:

$$H = \sum_{i} h(\boldsymbol{r}_i) + \frac{1}{2}\sum_{i,j} v_{ij} \tag{3.38}$$

where $h(\boldsymbol{r}_i)$ is the Hamiltonian for a single electron in a periodic potential, and $v_{ij} = \dfrac{e^2}{|\boldsymbol{r}_i - \boldsymbol{r}_j|}$.

In the Bloch representation, the second-quantized form of the Hamiltonian is

$$H = \sum_{k,\sigma} E_k C^{\dagger}_{k,\sigma} C_{k,\sigma} + \frac{1}{2} \sum_{k_1,k_2,k_1',k_2'} \sum_{\sigma,\sigma'} \langle k_1, k_2 | v | k_1', k_2' \rangle C^{\dagger}_{k_1,\sigma} C_{k_2,\sigma'} C_{k_2',\sigma'} C_{k_1',\sigma} \tag{3.39}$$

To better discuss electron correlation in narrow bands, it is preferable to adopt the Wannier representation. Using the representation transformation relations, the form of the Hamiltonian in the Wannier representation can be obtained

$$H = \sum_{i,j}\sum_{\sigma} T_{ij} C^{\dagger}_{i\sigma} C_{j\sigma} + \frac{1}{2} \sum_{i,j,l,m} \sum_{\sigma,\sigma'} \langle ij | v | lm \rangle C^{\dagger}_{i\sigma} C^{\dagger}_{j\sigma'} C_{m\sigma'} C_{l\sigma} \tag{3.40}$$

where $C^{\dagger}_{i\sigma}$ and $C_{i\sigma}$ are the creation and annihilation operators for spin-σ electrons. The single-particle term is

$$T_{ij} = \int a^{*}(\boldsymbol{r} - \boldsymbol{R}_i) h(\boldsymbol{r}) a(\boldsymbol{r} - \boldsymbol{R}_j) d\boldsymbol{r} = N^{-1} \sum_{k} e^{i\boldsymbol{k}\cdot(\boldsymbol{R}_i - \boldsymbol{R}_j)} E_{\boldsymbol{k}} \tag{3.41}$$

The interaction term

$$< ij | v | lm = e^2 \int \frac{a^{*}(\boldsymbol{r} - \boldsymbol{R}_i) a^{*}(\boldsymbol{r}' - \boldsymbol{R}_j) a(\boldsymbol{r} - \boldsymbol{R}_l) a(\boldsymbol{r}' - \boldsymbol{R}_m)}{|\boldsymbol{r} - \boldsymbol{r}'|} d\boldsymbol{r} d\boldsymbol{r}' \tag{3.42}$$

For narrow-band problems, the aforementioned interaction term can be simplified to a single-center integral form:

$$< ii|v|ii \geq e^2 \int \frac{a^*(\boldsymbol{r}-\boldsymbol{R}_i)a^*(\boldsymbol{r}'-\boldsymbol{R}_i)a(\boldsymbol{r}-\boldsymbol{R}_i)a(\boldsymbol{r}'-\boldsymbol{R}_i)}{|\boldsymbol{r}-\boldsymbol{r}'|} d\boldsymbol{r}d\boldsymbol{r}' \equiv U \quad (3.43)$$

reduced to

$$\frac{1}{2}U\sum_i\sum_{\sigma,\sigma'}C_{i\sigma}^\dagger C_{i\sigma'}^\dagger C_{i\sigma'}C_{i\sigma} = \frac{1}{2}U\sum_i\sum_\sigma n_{i\sigma}n_{i\bar{\sigma}} \quad (3.44)$$

Then the Hamiltonian of the Hubbard model is

$$H = \sum_{i,j}\sum_\sigma T_{ij}C_{i\sigma}^\dagger C_{j\sigma} + \frac{1}{2}U\sum_i\sum_\sigma n_{i\sigma}n_{i\bar{\sigma}} \quad (3.45)$$

The on-site Coulomb repulsion U for electrons with anti-parallel spins exerts a significant influence on the metal–insulator transition as well as on narrow-band magnetism.

When $j = i$, the single particle is

$$T_{ii} = N^{-1}\sum_{\boldsymbol{k}} E_{\boldsymbol{k}} \quad (3.46)$$

It represents the average energy of the band electrons. When $\boldsymbol{R}_j - \boldsymbol{R}_i$ is the nearest-neighbor lattice vector difference, $T_{ij} = T_1$, which is proportional to the bandwidth Δ in the TB approximation

When the bandwidth $\Delta = 0$, the Hamiltonian is written as

$$H = T_0\sum_{i,\sigma} n_{i\sigma} + \frac{1}{2}U\sum_i\sum_\sigma n_{i\sigma}n_{i\bar{\sigma}} \quad (3.47)$$

The single-particle Green's function is defined as

$$C_{ij}^\sigma(\omega) \equiv \ll C_{i\sigma} \big| C_{j\sigma}^\dagger >>_\omega \quad (3.48)$$

Using the Green's function, the exact solution for an isolated atomic system can be obtained, and its local density of states is given by

$$\rho_\sigma(\omega) = (1 - < n_{i\bar{\sigma}} >)\delta(\omega - T_0) + \langle n_{i\bar{\sigma}}\rangle\delta(\omega - T_0 - U) \quad (3.49)$$

When $\Delta = 0$ in the Hubbard model, there are two energy levels for electrons on each lattice site: T_0 and $T_0 + U$.

In the following, we focus on the analysis of strong correlation effects in narrow bands.[1] When the bandwidth $\Delta \neq 0$, the single-particle Green's function is

$$G_k^\sigma(\omega) = \frac{\omega - T_0 - U(1 - \langle n_{i\bar{\sigma}} \rangle)}{(\omega - E_k)(\omega - T_0 - U) + \langle n_{i\bar{\sigma}} \rangle U (T_0 - E_k)} = \frac{A_{k\sigma}^{(1)}}{\omega - E_{k\sigma}^{(1)}} + \frac{A_{k\sigma}^{(2)}}{\omega - E_{k\sigma}^{(2)}} \tag{3.50}$$

Then the local density of states is

$$\rho_\sigma(\omega) = \frac{i}{2\pi N} \sum_k \left\{ G_k^\sigma(\omega + i\eta) - G_k^\sigma(\omega - i\eta) \right\}, (\eta = +0) \tag{3.51}$$

Substituting Eq. (3.50) into the expression above, we obtain the following:

$$\rho_\sigma(\omega) = \frac{1}{N} \sum_k \left\{ A_{k\sigma}^{(1)} \delta\left(\omega - E_{k\sigma}^{(1)}\right) + A_{k\sigma}^{(2)} \delta\left(\omega - E_{k\sigma}^{(2)}\right) \right\} \tag{3.52}$$

Due to electron correlation, a single band splits into two subbands, such as $U = 0$, the Green's function and density of states, respectively, are as follows:

$$G_k^\sigma(\omega)\Big|_{U=0} = \frac{1}{\omega - E_k} \tag{3.53}$$

$$\rho_\sigma(\omega)\Big|_{U=0} = \frac{1}{N} \sum_k \delta(\omega - E_{k\sigma}) \tag{3.54}$$

Qualitative changes in both the band structure and density of states are illustrated in Figure 3.6a and b.

For the strong correlation case of $U \le \Delta$, we set

$$U \le |T_0 - E_k| \tag{3.55}$$

The Green's function and density of states are

$$G_k^\sigma(\omega) \approx \left\{ \frac{\langle n_{i\bar{\sigma}} \rangle}{\omega - T_0 - U - (E_k - T_0)\langle n_{i\bar{\sigma}} \rangle} + \frac{1 - \langle n_{i\bar{\sigma}} \rangle}{\omega - T_0 - (E_k - T_0)(1 - \langle n_{i\bar{\sigma}} \rangle)} \right\} \tag{3.56}$$

$$\rho_\sigma(\omega) \approx \frac{1}{N} \sum_k \{(1 - \langle n_{i\bar{\sigma}} \rangle) \delta\left[\omega - T_0 - (E_k - T_0)(1 - \langle n_{i\bar{\sigma}} \rangle)\right] + \langle n_{i\bar{\sigma}} \rangle \delta\left[\omega - T_0 - U - (E_k - T_0)\langle n_{i\bar{\sigma}} \rangle\right]\} \tag{3.57}$$

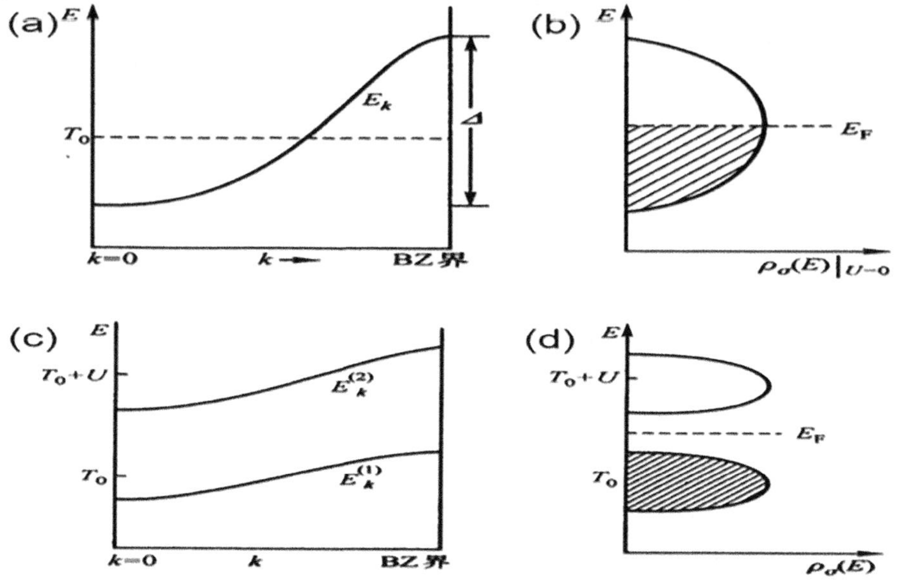

FIGURE 3.6 (a) and (b) The band structure and density of states for $U = 0$ case. (c) and (d) The band structure and density of states for $U \leq \Delta$ case.[1]

The two subbands are centered at energies T_0 and $T_0 + U$, the exact solution of the band structure and density of states can be referred to[6], here the schematic of the subband structure and the density of states is shown in Figure 3.6c and d.

It is important to note that when the wavefunctions of localized atomic orbitals on different lattice sites do not overlap, electrons in the crystal can only occupy the discrete energy levels T_0 and $T_0 + U$. When U is large, and the overlap between localized orbitals is small with $U \gg \Delta$, the T_0 and $T_0 + U$ levels from individual sites form two narrow, non-overlapping subbands. In the half-filled case ($n = 1$), the crystal remains an insulator. Conversely, when U is small and the orbital wavefunctions have significant overlap ($\Delta \gg U$), the two subbands centered at T_0 and $T_0 + U$ will hybridize and overlap broadly. In this regime, the crystal exhibits metallic behavior. Therefore, the ground-state property of a half-filled crystal, when electron correlations are accounted for, depends critically on the relative strength of U compared to the bandwidth (Δ). A metal–insulator transition occurs at a specific critical value of this ratio $(\Delta / U)_c$.

3.3.2 *T–J* Model

The discovery of high-temperature superconductivity in copper oxides in 1986 overturned the conventional superconductivity framework based on electron–phonon coupling. In their undoped state, these materials are antiferromagnetic Mott insulators. Upon doping with holes or electrons, the antiferromagnetic order is suppressed, giving rise to an anomalous pseudo-gap phase that eventually evolves into a d-wave superconductor at low temperatures.[7] This behavior cannot be explained by traditional band theory for weakly correlated electrons; its core physics lies in strong electron correlations. To capture this physics, theorists required a microscopic model that goes beyond BCS. While the full Hubbard model incorporates the on-site repulsion U, it is extremely difficult to solve. In the strong repulsion limit, the low-energy effective theory simplifies to the more refined *t–J* model.[8, 9] It eliminates the high-energy degrees of freedom (doubly occupied states) of the Hubbard model by directly imposing the "no double occupancy" constraint, while retaining the two key interactions: nearest-neighbor electron hopping and spin exchange. Consequently, the *t–J* model has become the central theoretical framework for studying the competition and coexistence of magnetism and superconductivity in doped Mott insulators.

The single-band Hubbard model is

$$H = -t\sum_{\langle ij\rangle,\sigma}\left(C_{i\sigma}^{\dagger}C_{j\sigma} + h.c.\right) + U\sum_{i} n_{i\uparrow}n_{i\downarrow} \tag{3.58}$$

with $\langle ij\rangle$ represents the nearest neighbor, t is the hopping term. For $U = 0$, the system is in the non-interacting limit, described by a TB band with a capacity of $2N$ electrons. At half-filling ($n = 1$ electron per site), this band is exactly half-full. In contrast, the regime with $U/t >> 1$ is known as the strongly correlated limit of the Hubbard model.

The single-band Hamiltonian can be written as $H = T + V$, with $T = -t\sum_{\langle ij\rangle,\sigma}\left(C_{i\sigma}^{\dagger}C_{j\sigma} + h.c.\right)$ and $V = U\sum_{i} n_{i\uparrow}n_{i\downarrow}$. By employing a standard identity $C_{i\sigma} = C_{i\sigma}\left[\left(1 - n_{i\bar{\sigma}}\right) + n_{i\bar{\sigma}}\right]$, the kinetic energy term T can be divided into

$$T = -t\sum_{\langle ij\rangle,\sigma}\left[\left(1 - n_{i\bar{\sigma}}\right) + n_{i\bar{\sigma}}\right]C_{i\sigma}^{\dagger}C_{j\sigma}\left[\left(1 - n_{j\bar{\sigma}}\right) + n_{j\bar{\sigma}}\right] + h.c. = T_h + T_d + T_{\text{mix}} \tag{3.59}$$

with the hopping term of an electron from a singly occupied site j to an empty site i:

$$T_h = -t\sum_{\langle ij\rangle,\sigma}\left[\left(1 - n_{i\bar{\sigma}}\right)C_{i\sigma}^{\dagger}C_{j\sigma}\left(1 - n_{j\bar{\sigma}}\right) + h.c\right] \tag{3.60}$$

This hopping process takes place within the lower Hubbard band. The term describing the motion of an electron via an empty site in this band is simply called the "lower-band hole" term.

The hopping term of an electron from a doubly occupied site j to a single occupied site i:

$$T_d = -t\sum_{\langle ij\rangle,\sigma}\left[n_{i\bar{\sigma}}C_{i\sigma}^{\dagger}C_{j\sigma}n_{j\bar{\sigma}} + h.c\right] \tag{3.61}$$

This process transfers the second electron from site j in the upper band to site i, thereby describing the hopping of a doubly occupied site within the upper band.

The mixed hopping term is

$$T_{\text{mix}} = -t\sum_{\langle ij\rangle,\sigma}\left[\left(1-n_{i\bar{\sigma}}\right)C_{i\sigma}^{\dagger}C_{j\sigma}n_{j\bar{\sigma}} + n_{i\bar{\sigma}}C_{i\sigma}^{\dagger}C_{j\sigma}\left(1-n_{j\bar{\sigma}}\right) + h.c\right] \tag{3.62}$$

The T_{mix} term describes an interband transition that hybridizes the "upper" and "lower" bands. In the strong correlation regime ($U/t >> 1$) with less than half-filling ($n < 1$), although the upper band is prohibitively high in energy to be physically occupied, it can act as a virtual intermediate state. Through such virtual processes, it mediates effective interactions that significantly influence electron motion in the lower band. Consequently, the T_{mix} term plays a vital role in the t–J model.

The projection operator is defined as

$$P = \prod_{i=1}^{n}\left(1-n_{i\uparrow}n_{i\downarrow}\right) \tag{3.63}$$

$Q = 1-P$ represents another projection operator, then the mixing term can be reduced as

$$T_{\text{mix}} = -PT_{\text{mix}}Q + QT_{\text{mix}}P \tag{3.64}$$

After removing the first-order contributions from the hopping term T_{mix}, the effective Hamiltonian for the Hubbard model, correct to second order in perturbation theory, takes the form:

$$H_{\text{eff}} = T_h + T_d + V + \frac{1}{2}\left[T_{\text{mix}}, S\right] \tag{3.65}$$

with the generating function of the transformation,

$$S = \frac{PT_{\text{mix}}Q - QT_{\text{mix}}P}{U} \tag{3.66}$$

Consequently, the low-energy effective Hamiltonian for the single-band Hubbard model in the strongly correlated limit is derived as follows:

$$H_{\text{eff}} = T_h - \frac{1}{U}\left(PT_{\text{mix}}Q\right)\left(QT_{\text{mix}}P\right) \tag{3.67}$$

Substituting the explicit form of the projection operator, we obtain the expression for the Hamiltonian

$$H = -t\sum_{\langle ij\rangle,\sigma}\left(\hat{C}_{i\sigma}^{\dagger}\hat{C}_{j\sigma} + h.c.\right) + J\sum_{i,j}\left(\hat{S}_i\cdot\hat{S}_j - \frac{1}{4}n_in_j\right) \tag{3.68}$$

with

$$\hat{C}_{i\sigma}^{\dagger} = (1 - \eta_{i\bar{\sigma}}) C_{i\sigma}^{\dagger}, \hat{C}_{j\sigma} = C_{j\sigma}(1 - n_{j\bar{\sigma}}) \tag{3.69}$$

and spin operator

$$\hat{S}_i^z = \frac{1}{2}(C_{i\uparrow}^{\dagger}C_{i\uparrow} - C_{i\downarrow}^{\dagger}C_{i\downarrow}), \hat{S}_i^+ = C_{i\uparrow}^{\dagger}C_{i\downarrow}, \hat{S}_i^- = C_{i\downarrow}^{\dagger}C_{i\uparrow} \tag{3.70}$$

where $n_i(n_j)$ represents the particle number density on-site $i(j)$. $J = 4t^2/U > 0$ describes the antiferromagnetic exchange interaction between nearest-neighbor spins $\hat{S}_i$ and $\hat{S}_i$, $\hat{C}_{i\sigma}^{\dagger}$ and $\hat{C}_{j\sigma}$ are operators in the P-subspace.

In the strongly correlated and nearly half-filled regime, the t–J model takes the form:

$$H = -(t\delta)\sum_{\langle ij\rangle,\sigma}(C_{i\sigma}^{\dagger}C_{j\sigma} + h.c.) + J\sum_{i,j}\left(\hat{S}_i \cdot \hat{S}_j - \frac{1}{4}n_i n_j\right) \tag{3.71}$$

with $\hat{S}_i = \frac{1}{2}\sum_{\alpha,\beta} C_{i\alpha}^{\dagger}(\hat{\sigma})C_{i\beta}^{\dagger}$, α, β is the spin index, $\hat{\sigma}$ is the vector Pauli matrix. $\delta = 1 - n$ represents the average hole density per site.

The *t–J* model is widely used in high-temperature superconductor research for its concise description of the low-energy excitations and competing interactions central to the materials in-plane behavior. This has made it one of the most studied models in condensed matter physics.

3.4 QUANTUM HALL EFFECT

The classical Hall effect was discovered by the American physicist Edwin Hall in 1879. It describes how, when an electric current passes through a conductor or semiconductor placed in a perpendicular magnetic field, a measurable transverse voltage difference, known as the Hall voltage, is generated in the direction perpendicular to both the current and the magnetic field. The Hall voltage is

$$V_H = R_H \cdot \frac{IB}{d} \tag{3.72}$$

with R_H is the Hall coefficient, I and B, respectively, are the electric current and magnetic field strength, d is the thickness.

The Hall resistivity (transverse resistivity) ρ_{xy} is given by

$$\rho_{xy} = R_H B = \frac{V_H}{I} \tag{3.73}$$

Under the classical model, the Hall resistivity is proportional to the magnetic field. A key prediction of the classical Hall effect is the linear dependence of the Hall resistance on the magnetic field.

A century later, in a 2D electron gas at ultra-low temperatures (liquid helium temperatures) and under strong magnetic fields, the German physicist Klaus von Klitzing discovered the integer quantum Hall effect (IQHE)[10,11] in 1980. The Hall conductance exhibits highly precise quantized plateaus, $\sigma_{xy} = ve / h$, where v is an integer. This discovery, due to its extraordinary measurement precision (on the order of 10^{-9}), is used as an international resistance standard. Klaus von Klitzing was awarded the Nobel Prize in Physics in 1985 for this work.

Just two years later, using purer samples and even stronger magnetic fields, Daniel C. Tsui, Horst L. Störmer, and Arthur C. Gossard discovered the fractional quantum Hall effect (FQHE), characterized by fractional v values (such as $1/3, 2/5$). This phenomenon could not be explained by a single-electron picture, marking the birth of a novel quantum state of matter driven by strong electron–electron correlations. Robert B. Laughlin subsequently proposed the famous Laughlin wavefunction theory, which predicted the existence of quasi-particles carrying fractional electric charge. The three shared the Nobel Prize in Physics in 1998.

Research on the quantum Hall effects has given rise to entirely new physical concepts such as "topological order" and "topological insulators." It continues to attract significant attention due to the anyons it hosts, quasi-particles with statistics intermediate between bosons and fermions, and their potential application in topological quantum computing. This chapter will systematically elucidate the physical imagery, theoretical explanations, and recent advances of the quantum Hall effect.

The IQHE reveals that the relationship between the Hall resistance R_H and the magnetic field strength B is characterized by a series of plateaus superimposed on an overall linear trend, as shown in Figure 3.7a.

The resistance at these plateaus is quantized as

$$R_H = \frac{h}{ie^2} \tag{3.74}$$

where i is a positive integer, h is Planck's constant, and e is the electron charge. When $i = 1$, the quantized Hall resistance h / e^2 is known as the von Klitzing constant. Since 1990, the International Committee for Weights and Measures (CIPM) has adopted the quantum Hall resistance as the primary standard for resistance metrology. The quantum Hall effect provides an experimental approach within condensed matter physics that is independent of quantum electrodynamics, allowing for the determination of the fundamental constants of nature $\alpha = e^2 / (4\pi \epsilon_0 c\hbar), c$ is the speed of light in vacuum.

The physical essence of the IQHE can be understood through the Landau levels formed by electrons under a strong magnetic field and their degeneracy. In a perpendicular, uniform, and strong magnetic field B, the Hamiltonian of a single electron (in the symmetric gauge) can be written as follows:

$$H = \frac{1}{2m}(\hat{p} + e\mathbf{A})^2 \tag{3.75}$$

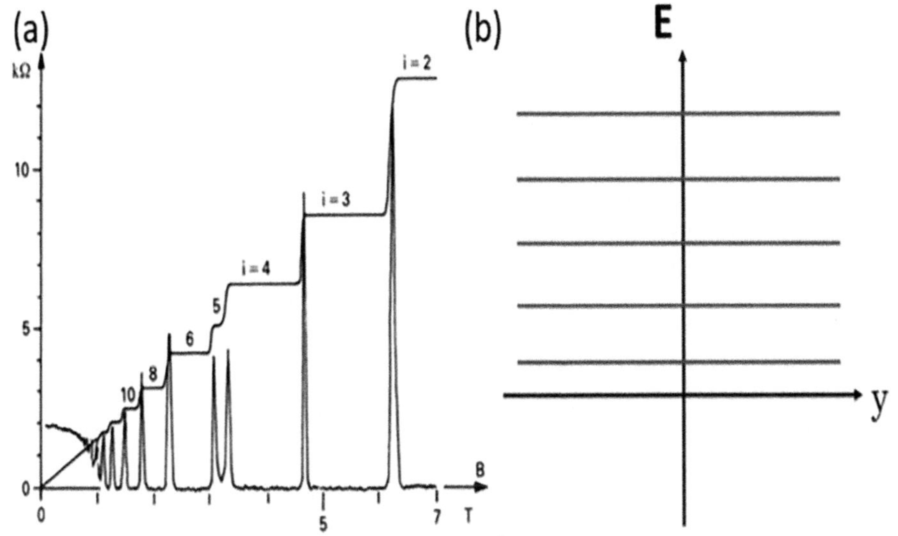

FIGURE 3.7 (a) The IQHE. (b) The band structure.

where m is the effective mass of the electron, $\hat{p}$ is the momentum operator, e is the absolute value of the electron charge, and $\boldsymbol{A}$ is the magnetic vector potential (satisfying $\nabla \times \boldsymbol{A} = B\hat{z}$).

Solving this Hamiltonian yields a series of discrete energy levels, known as Landau levels:

$$E_n = \hbar\omega_c\left(n + \frac{1}{2}\right), n = 0,1,2,\ldots \tag{3.76}$$

with $\omega_c = eB/m$ is the cyclotron frequency. The energy level spacing is determined by the magnetic field B, and the degeneracy of the Landau levels is proportional to B. As the magnetic field varies, a quantized plateau in the Hall resistance emerges when exactly i Landau levels are completely filled with electrons.

Each Landau level possesses a high degree of degeneracy. In a 2D system, this degeneracy N is proportional to both the magnetic field strength B and the sample area S:

$$N = \frac{eB}{h}S = \frac{BS}{\Phi_0} \tag{3.77}$$

where $\Phi_0 = h/e$ is the magnetic flux quantum. This implies that the maximum number of electrons each Landau level can accommodate is determined by the total magnetic flux penetrating the sample.

The underlying physical mechanism can be understood as follows: disorder present in real samples broadens the Landau levels and creates localized states within the energy gaps. When the Fermi level E_F lies within a region of localized states, these states do not contribute to conduction, while the occupation of the extended states (corresponding to the centers of the Landau levels) remains unchanged. Consequently, even as the magnetic field B changes (causing the filling factor i to vary), the Hall conductance remains constant as long as E_F moves through the localized states, thereby forming a plateau. The quantum Hall effect can only be realized under conditions of strong magnetic fields and low temperatures.

In the bulk of a material, under the influence of a magnetic field, electrons undergo closed circular motion and remain localized around the center of their orbits, contributing little to conduction. However, at the material's edge, the presence of the boundary prevents electrons from completing full circular paths. Instead, the motion of edge electrons consists of repeated semicircular segments, which combine to form propagating edge states. The number of these edge states determines the integer index in the IQHE. Consequently, the Hall conductance of an IQHE is determined by a topological invariant known as the Chern number. The Chern number is an integer that describes the global geometric property of the electron wavefunction across the entire momentum space (the BZ) through the integral of the Berry curvature. Theoretically, it is shown that the Hall conductance satisfies:

$$\sigma_{xy} = C\frac{e}{h} \tag{3.78}$$

where C is the Chern number of the system. As long as the Fermi level lies within an energy gap, the Chern number remains an integer invariant under weak changes in microscopic details, such as disorder or sample geometry, thereby directly accounting for the precise quantization of the Hall conductance.

Using the Kubo formula, the conductivity can be expressed as follows:

$$\begin{aligned}\sigma_{xy} &= \frac{ie}{\hbar}\sum_{\alpha}\int_{T^2}\frac{d^2k}{(2\pi)^2}\left(\left\langle \partial_y u_k^\alpha \mid \partial_x u_k^\alpha \right> - \left\langle \partial_x u_k^\alpha \middle| \partial_y u_k^\alpha \right\rangle\right) \\ &= \frac{e}{2\pi\hbar}\sum_{\alpha} C_\alpha\end{aligned} \tag{3.79}$$

This result, which expresses the quantized Hall conductance as a topological invariant (the Chern number), is famously known as the TKNN formula. Where the Chern number is

$$C_\alpha = \frac{i}{2\pi}\int_{T^2} d^2k\left(\left\langle \partial_y u_k^\alpha \mid \partial_x u_k^\alpha \right> - \left\langle \partial_x u_k^\alpha \middle| \partial_y u_k^\alpha \right\rangle\right) \tag{3.80}$$

When $\Sigma_\alpha C_\alpha \neq 0$, the system exhibits the characteristic IQHE and is referred to as a Chern insulator. A key property of a Chern insulator is the breaking of time-reversal symmetry.

Unlike the IQHE, which is explained by single-particle band topology and disorder-induced localization, the FQHE is a phenomenon observed in 2D electron systems under extremely low temperatures and strong magnetic fields. It occurs when the Landau level filling factor takes on fractional values (e.g., $v = 1/3, 2/5$), leading to fractional quantized plateaus in the Hall conductivity ($\sigma_{xy} = (p/q) \cdot e^2/h$). The FQHE is fundamentally rooted in strong electron–electron correlations, defying explanation within any single-particle framework. By progressively increasing the external magnetic field in suitable materials, a transition from the IQHE to the FQHE can be realized. This corresponds to the evolution of a 2D electron gas under magnetic field tuning: from a topological phase dominated by single-particle Landau level quantization and disorder-induced localization, to a phase governed by strong electron–electron correlations and many-body entanglement. In other words, the system undergoes a fundamental transition from a weakly interacting regime to a strongly interacting one.

The quantum spin Hall effect (QSHE) is a topological phase of matter in 2D systems that exhibits quantized spin Hall conductance while maintaining zero charge Hall conductance. Unlike the IQHE, the QSHE occurs in the absence of an external magnetic field and fundamentally relies on the protection of time-reversal symmetry. In contrast, the FQHE is a strongly correlated topological state that emerges under strong magnetic fields. While both the QSHE and FQHE involve electronic correlations, the QSHE originates from band inversion driven by strong spin–orbit coupling (e.g., in HgTe quantum wells), and can largely be described within a non-interacting topological picture. In contrast, the FQHE is fundamentally a many-body phenomenon that arises in high magnetic fields and explicitly breaks time-reversal symmetry.

The QSHE arises from strong spin–orbit coupling, which acts as a momentum-dependent effective magnetic field. This field points in opposite directions for electrons with opposite spins and momenta, resulting in a spin-dependent force that drives the formation of one-dimensional channels at the sample edges. Within these channels, electron motion is locked to its spin orientation: spin-up electrons propagate in one direction, while spin-down electrons move in the opposite direction, giving rise to a pure spin current without net charge flow. These helical edge states are protected by time-reversal symmetry. Because backscattering would require a spin flip, a process forbidden as long as time-reversal symmetry is preserved, the edge conduction remains robust against non-magnetic disorder. Materials exhibiting the QSHE are referred to as topological insulators.[12,13]

Topological insulators are characterized by an insulating bulk and conducting boundaries, where the latter are topologically protected and robust against non-magnetic disorder. In 2007, the Molenkamp group first observed the quantum spin Hall effect in HgTe/CdTe quantum wells.[14] By precisely controlling the quantum well thickness, band inversion was achieved, driving the system into a topologically nontrivial phase beyond a critical thickness. The edge transport exhibited a quantized conductance plateau, demonstrating robustness against non-magnetic disorder.

REFERENCES

[1] Li, Z. Z. *Solid State Theory.* , Beijing: Higher Education Press.

[2] Huang, K. *Solid State Physics.* Beijing: Peking University Press.

[3] Reif-Acherman, S. Studies on the temperature dependence of electric conductivity for metals in the nineteenth century: A neglected chapter in the history of superconductivity. *Revista Brasileira de Ensino de Física* 33, 2601 (2011).

[4] Zhang, L. Y. *Beyond Freedom: The Marvel of Superconductors*. Beijing: Science Press (2005).
[5] Wu, H. S., Guan, W. Y., Li, H. C. *Superconductivity: Type II Superconductors and Weakly Linked Superconductors*. Beijing: Peking University Press (2014).
[6] Hubbard, J. Electron correlations in narrow energy bands. *Proceedings of the Royal Society (London)* A276, 238 (1963).
[7] Bednorz, J. G., Müller, K. A. Z. Possible high T_c superconductivity in the Ba–La–Cu–O system. *Physica B* 64 (2), 189–193 (1986).
[8] Zhang, F. C., Rice, T. M. Effective Hamiltonian for the superconducting Cu oxides. *Physical Review B* 37 (7), 3759 (1988).
[9] Anderson, P. W. The resonating valence bond state in La_2CuO_4 and superconductivity. *Science* 235 (4793), 1196–1198 (1987).
[10] von Klitzing, K. The quantized hall effect. *Reviews of Modern Physics* 58, 519 (1986).
[11] Girvin, S. M. The Quantum hall effect: Novel excitations and broken symmetries. In: Comtet, A., Joliecure, T., Ouvry, S., David, F. (eds.), *Aspects topologiques de la physique en base dimension. Topological aspects of low dimensional systems*. Berlin: Springer (1999).
[12] Chen, Y., et al. Topological carbon materials: A new perspective. *Physics Reports* 868, 1–32 (2020).
[13] Sharma, V. et al. Chemical ordering and electronic properties of lone pair chalcogenide semiconductors. *Progress in Solid State Chemistry* 54, 31–44 (2019).
[14] Felser, C., Qi, X. L. Topological insulators. *MRS Bulletin* 39, 843–846 (2014).

Part II

Major Systems of Low-Dimensional Quantum Materials

4 Low-Dimensional Topological Materials

Dan Mu, Long Ren, and Jianxin Zhong

4.1 BASIC OVERVIEW

Topological materials

In condensed matter physics, the understanding of phases and phase transitions has traditionally been based on Landau's symmetry-breaking paradigm, in which distinct phases are classified by the symmetries of their ground states and characterized by local order parameters. However, the discovery of the quantum Hall effect revealed that some quantum states cannot be fully understood within this framework. Instead, they are distinguished by topological invariants that encode the global structure of electronic wavefunctions in momentum space, thereby opening a new route for classifying phases of matter. Against this background, the modern field of topological materials emerged from studies of quantum Hall physics, time-reversal-invariant band topology, and the quantum spin Hall effect.

The modern field of topological materials emerged from theoretical studies of time-reversal-invariant band topology and the quantum spin Hall effect. Kane and Mele first proposed that graphene with spin–orbit coupling (SOC) could host a two-dimensional (2D) topological insulating phase characterized by helical edge states protected by time-reversal symmetry.[2, 3] This idea was subsequently generalized to semiconductor quantum wells, where a topological phase transition driven by band inversion was predicted in HgTe/CdTe heterostructures.[1] These theoretical developments established the fundamental framework of topological band theory, which was later systematically summarized in several influential reviews, as shown in Figures 4.1–4.4.[4,5]

Experimentally, the first realization of a 2D topological insulator was achieved in HgTe/CdTe quantum wells, where transport measurements revealed quantized edge conductance associated with the quantum spin Hall state.[6] Shortly afterward, angle-resolved photoemission spectroscopy (ARPES) experiments identified three-dimensional (3D) topological insulators in the Bi_2Se_3 family compounds, demonstrating the existence of gapless Dirac surface states protected by the topology of the bulk electronic bands in Figure 4.5.[7,8] These discoveries established topological insulators as a new quantum phase of matter distinct from conventional band insulators.

The conceptual foundation of topological phases can be traced back to the discovery of the quantum Hall effect in 2D electron systems. In this phenomenon, the Hall conductivity exhibits precisely quantized plateaus that are extraordinarily robust

DOI: 10.1201/9781003655305-6

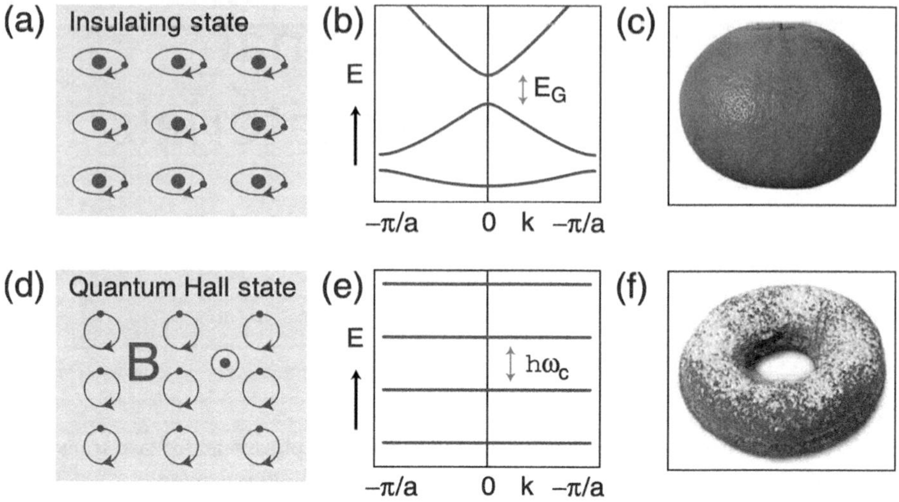

FIGURE 4.1 States of matter. (a–c) The insulating state. (a) An atomic insulator. (b) A simple model of an insulating band structure. (d–f) The quantum Hall state. (d) The cyclotron motion of electrons. (e) The Landau levels, which may be viewed as a band structure. (c) and (f) Two surfaces which differ in their genus, g. (c) $g=0$ for the sphere and (f) $g=1$ for the donut. The Chern number n, which distinguishes the two states, is a topological invariant analogous to the genus.[1]

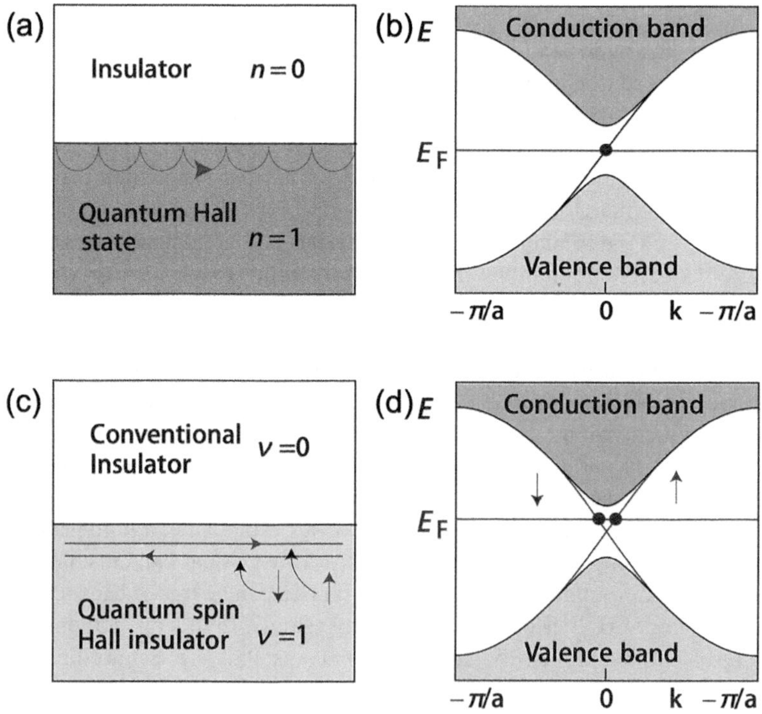

(Continued)

FIGURE 4.2 (CONTINUED) The interface between a quantum Hall state and an insulator has a chiral edge mode and edge states in the quantum spin Hall insulator (QSHI). (a) The skipping cyclotron orbits. (b) The electronic structure of a semi-infinite strip described by the Haldane model. A single edge state connects the valence band to the conduction band. (c) The interface between a QSHI and an ordinary insulator. (d) The edge state dispersion in the graphene model, in which up and down spins propagate in opposite directions.[1]

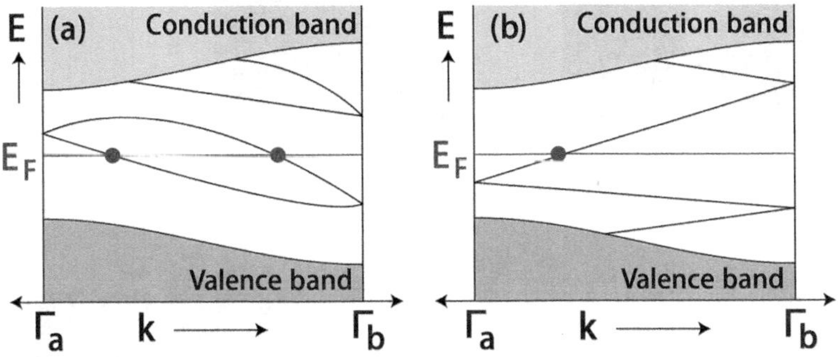

FIGURE 4.3 Electronic dispersion between two boundary Kramers degenerate points $\Gamma_\alpha = 0$ and $\Gamma_b = \pi / a$. In the number of surface states crossing the Fermi energy E_F is even, whereas in b it is odd. An odd number of crossings leads to topologically protected metallic boundary states.[1]

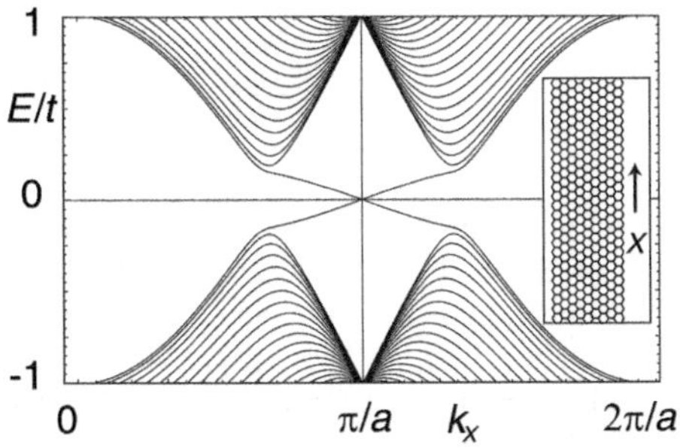

FIGURE 4.4 One-dimensional (1D) energy bands for a strip of graphene (shown in the inset). The bands crossing the gap are spin-filtered edge states.[2]

against variations in disorder and microscopic material parameters.[9] Importantly, transitions between Hall plateaus occur without any change in the ground-state symmetry, indicating that these phases cannot be described within the traditional Landau symmetry-breaking framework.

The topological nature of the quantum Hall state was rigorously established by Thouless, Kohmoto, Nightingale, and den Nijs (TKNN), who demonstrated that the quantized Hall conductance is determined by an integer topological invariant, now known as the TKNN invariant or the first Chern number of the occupied electronic

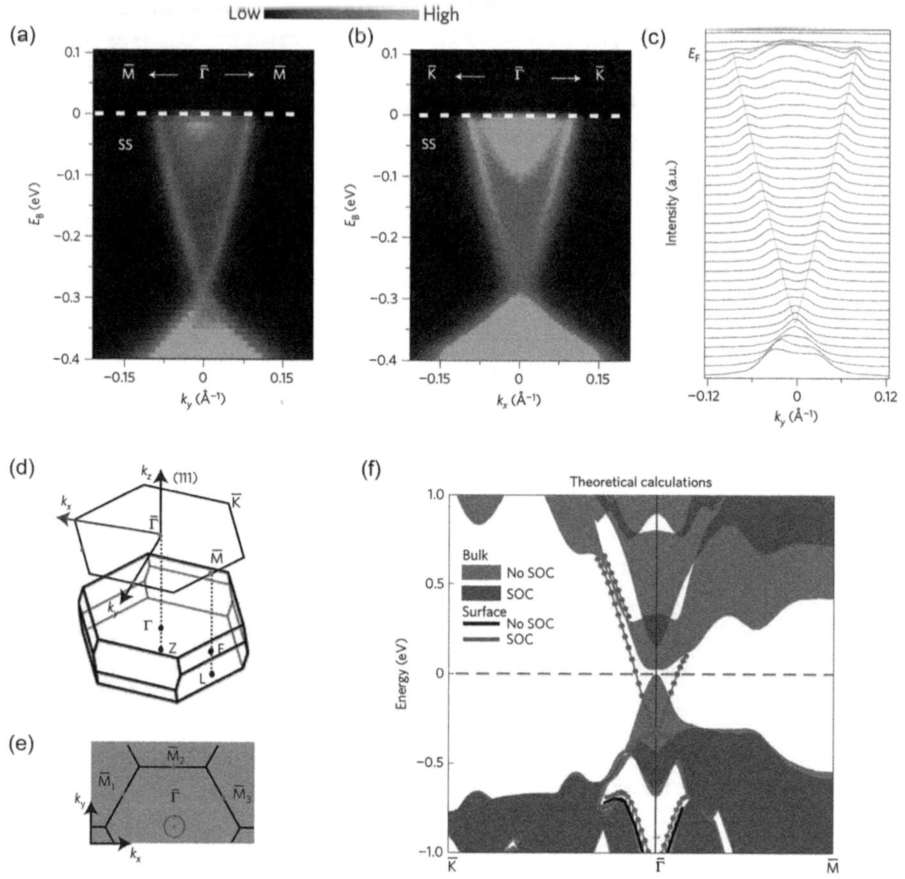

FIGURE 4.5 Strong SOC generates a single surface-state Dirac cone. (a and b) High-resolution ARPES spectra of $Bi_2Se_3(111)$ measured with 22 eV photons along the $\Gamma - M$ and $\Gamma - K$ directions. (c) Momentum distribution curves showing that the two surface bands merge into a single Dirac point at Γ. The nearly isotropic V-shaped bands define a Dirac cone, whereas the broad U-shaped feature arises from the conduction band bottom. (d) Schematic of the bulk and projected surface Brillouin zones. (e) The surface Fermi surface is a single ring centered at Γ when the chemical potential lies inside the bulk gap. The single Fermi-level crossing between the two time-reversal invariant momenta (TRIMs), Γ and M, implies a π Berry phase. (f) Local density approximation (LDA) band structure with projected bulk bands shown as shaded regions. A gapless surface band appears only when SOC is included, whereas no pure surface state is present in the gap without SOC.[7]

bands.[10] Physically, this invariant corresponds to the integral of the Berry curvature over the entire Brillouin zone and determines the Hall conductivity through $\sigma_{xy} = \nu \frac{e^2}{h}$, where ν is the Chern number. The robustness of the quantized Hall response therefore reflects the global topology of the electronic wavefunctions rather than microscopic details of the system.

From a broader perspective, topology enters condensed matter systems through the geometric properties of Bloch wavefunctions in momentum space. In a gapped crystalline system, the occupied electronic states form a smooth manifold across the Brillouin zone. The geometric structure of these states is captured by the Berry connection and Berry curvature, whose integrals define global topological invariants that remain unchanged under continuous deformations of the Hamiltonian as long as the bulk band gap remains open.[11]

An intuitive analogy arises from topology in mathematics, where closed surfaces are classified according to their genus, an integer invariant that counts the number of holes in the surface. Objects sharing the same genus can be smoothly deformed into one another, whereas surfaces with different genus cannot be continuously connected without introducing singular transformations. Similarly, topological invariants classify quantum phases into distinct equivalence classes that cannot be adiabatically connected without closing the energy gap.

Unlike quantum Hall systems, which require strong external magnetic fields to generate Landau-level quantization, topological insulators realize nontrivial topological phases in the absence of magnetic fields, with protection guaranteed by time-reversal symmetry. In these materials, strong spin–orbit coupling plays a crucial role by inducing band inversion between electronic states of different orbital character near the Fermi level, thereby driving the system into a topologically nontrivial regime. The resulting band topology is characterized by a Z_2 topological invariant, which determines whether a system is a trivial or nontrivial topological insulator.[1,4]

Over the past decade, the concept of topological band theory has expanded far beyond topological insulators to encompass a wide range of topological quantum materials, including topological semimetals, topological superconductors, and magnetic topological systems. For example, Dirac and Weyl semimetals host symmetry-protected band crossings that give rise to relativistic quasiparticles and exotic surface states such as Fermi arcs.[12] Topological superconductors are predicted to host Majorana bound states with potential applications in fault-tolerant quantum computation,[13] while magnetic topological materials can exhibit quantized anomalous Hall effects without external magnetic fields.[14]

Large-scale theoretical searches and materials databases have further revealed that topological electronic structures are far more common in crystalline solids than originally expected.[15,16] In particular, many experimentally accessible topological phases are realized in reduced-dimensional systems such as thin films, 2D materials, and 1D nanostructures.

These systems are collectively referred to as low-dimensional topological materials. In such systems, nontrivial topological invariants defined in reduced momentum space give rise to protected boundary states localized at edges or ends of the material. Reduced dimensionality often enhances quantum confinement, SOC effects, and symmetry constraints, making low-dimensional systems ideal platforms for realizing and manipulating topological quantum states.

Based on their electronic structures and physical properties, low-dimensional topological materials can be broadly classified into several major categories, including topological insulators, topological semimetals, topological semiconductors, topological superconductors, magnetic topological materials, and topological photonic systems, which will be discussed in detail in the following sections.

4.2 MAIN TYPES OF LOW-DIMENSIONAL TOPOLOGICAL MATERIALS

Building upon the general framework of topological band theory discussed above, low-dimensional topological materials can be classified according to the nature of their bulk topology and the corresponding protected boundary excitations. In reduced-dimensional systems, quantum confinement, symmetry constraints, and enhanced SOC often stabilize topological phases that are difficult to access in bulk crystals. As a result, 1D chains, 2D crystals, thin films, and artificially engineered lattices have become central platforms for investigating topological quantum matter. Based on their electronic structures and physical properties, low-dimensional topological materials can be broadly categorized into topological insulators, topological semimetals, topological semiconductors, topological superconductors, magnetic topological materials, and topological photonic materials.[4,5,12,13]

4.2.1 Low-Dimensional Topological Insulators

Low-dimensional topological insulators are the most representative and extensively studied class of topological materials. These systems are characterized by an insulating bulk and symmetry-protected boundary states. Their modern theoretical foundation was established by the Kane–Mele model, which introduced the quantum spin Hall effect in graphene and demonstrated that SOC can drive a 2D system into a nontrivial Z_2 topological phase.[2,3] This idea was subsequently generalized in the Bernevig–Hughes–Zhang model for HgTe/CdTe quantum wells, where band inversion was shown to induce a topological phase transition.[1]

The first decisive experimental realization of a 2D topological insulator was achieved in HgTe quantum wells, where quantized edge transport consistent with helical edge states was observed, as shown in Figure 4.6.[6] Similar topological behavior was later reported in InAs/GaSb quantum wells, which provided another important semiconductor platform for electrically tunable quantum spin Hall physics.[18] The field then expanded from conventional quantum wells to atomically thin materials. Monolayer 1T′-WTe_2 became a landmark system because both transport and spectroscopic studies established it as a robust van der Waals quantum spin Hall material, as illustrated in Figure 4.7.[17,19]

Recent work has increasingly focused on large-gap and device-compatible 2D topological insulators. Bismuthene on SiC has attracted broad interest as a candidate high-temperature quantum spin Hall material because of its large topological gap in Figure 4.8.[20] More recent work has explored graphene intercalation as a route to preserve and engineer its topological properties in realistic environments.[21] In parallel, elevated-temperature quantum spin Hall behavior has been pursued in III–V semiconductor systems, suggesting that low-dimensional topological insulators are gradually moving from low-temperature proof-of-principle platforms toward materials with greater practical potential.[22]

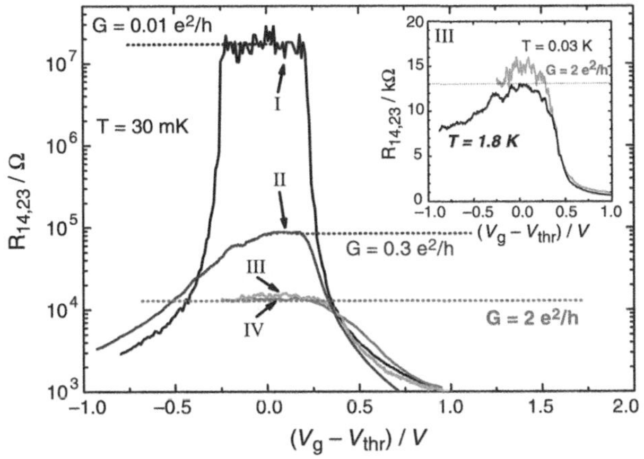

FIGURE 4.6 The longitudinal four-terminal resistance, $R_{14,23,}$ of various normal (d = 5.5 nm) (I) and inverted (d = 7.3 nm) (II, III, and IV) QW structures as a function of the gate voltage measured for B = 0 T at T = 30 mK. The device sizes are (20.0 × 13.3) mm^2 for devices I and II, (1.0 × 1.0) mm^2 for device III, and (1.0 × 0.5) mm^2 for device IV. The inset shows $R_{14,23}(V_g)$ of two samples from the same wafer, having the same device size (III) at 30 mK (III) and 1.8 K (I) on a linear scale.[6]

4.2.2 Low-Dimensional Topological Semimetals

Low-dimensional topological semimetals are distinguished by symmetry-protected band crossings rather than a full insulating gap. In these systems, conduction and valence bands intersect at discrete points or along lines in momentum space, producing quasiparticles that resemble relativistic Dirac or Weyl fermions.[12,23] As schematically inlustrated in Fig 4.9, different symmetry-breaking terms can transform a Dirac semimetal into magnetic semiconductor, Wely semimetal, or line-node semimetal phases.[24] Their topological nature is associated with Berry-curvature singularities and manifests experimentally through unusual transport responses and protected boundary states such as Fermi arcs (Figure 4.9).[12]

Experimentally, Dirac semimetal phases were first established in Na_3Bi and Cd_3As_2, where ARPES revealed linear bulk dispersions characteristic of Dirac fermions.[23,25] Shortly afterward, Weyl nodes and topological Fermi-arc surface states were observed in TaAs-family compounds, firmly establishing Weyl semimetals as a new class of topological quantum materials.[26,27] Although many of these canonical examples are 3D, their thin-film, surface, and layered derivatives provide important low-dimensional platforms in which topology can be tuned by thickness, strain, symmetry breaking, or interfacial coupling.

A major recent development in this area is the rise of kagome metals as low-dimensional topological semimetal platforms. In kagome lattices, geometric frustration and lattice symmetry naturally generate Dirac bands, flat bands, and van Hove singularities, leading to a rich interplay between topology and electronic correlation. The AV_3Sb_5 family, in particular, has become a central example because it combines

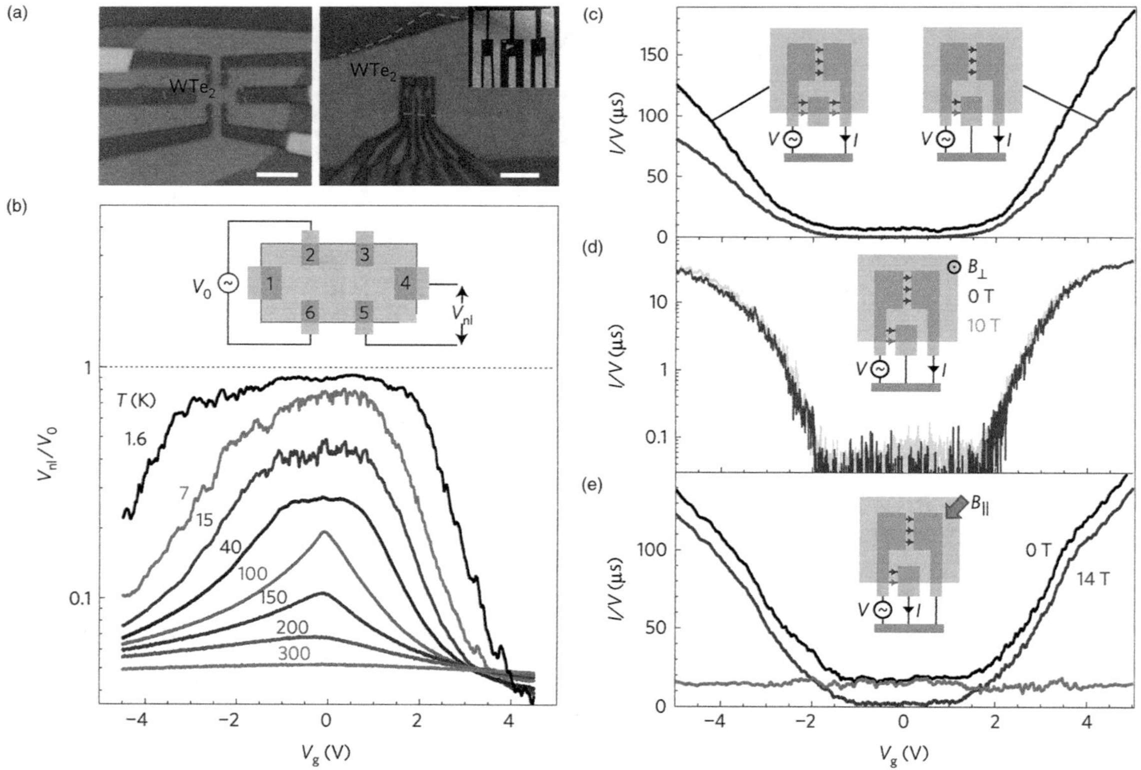

FIGURE 4.7 (a) Optical images of monolayer devices MW_2 and MW_3. Dashed lines mark the transferred monolayer sheets. Scale bars, 5 μm. Inset: AFM image of the electrodes. (b) Nonlocal transport measurements on MW_2 in the configuration shown in the inset (V_0 = 100 μV, 11.3 Hz). (c–e) Transport measurements on MW_3. (c) Two-terminal conductance between the outer contacts (up) and I/V with the center contact grounded (down) at 10 K. Insets indicate edge (lighter color) and bulk (darker color) current paths. (d) Same as the the bottom line trace in c, measured at 1.6 K on a logarithmic scale, showing only a weak response to a perpendicular magnetic field of 10 T. (e) Effect of an in-plane magnetic field ($B_{||}$=14 T) on I/V between adjacent contacts at 10 K; the horizontal line trace shows the magnitude of the suppression.[17]

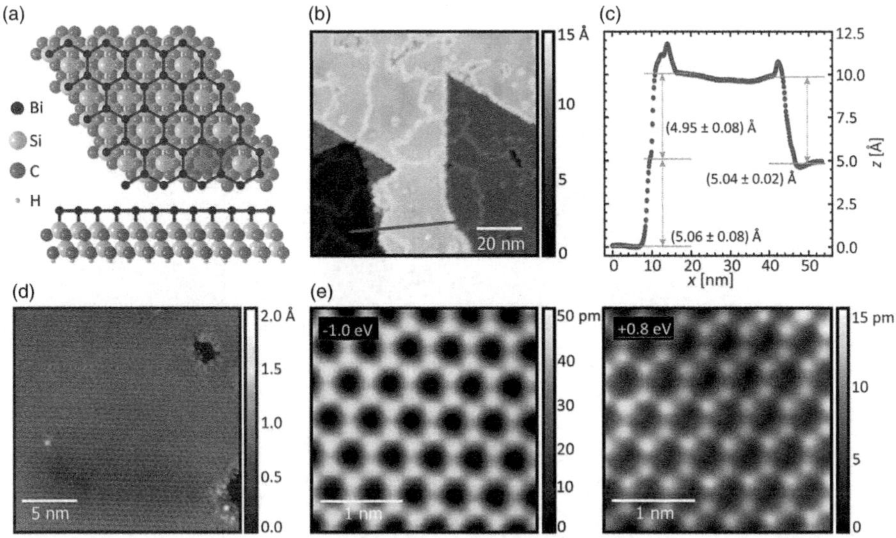

FIGURE 4.8 Bismuthene on SiC (0001) structural model. (a) Sketch of a bismuthene layer placed on the threefold-symmetric SiC (0001) substrate in $\sqrt{3} \times \sqrt{3}$ R30° commensurate registry. (b) Topographic scanning tunneling microscopy (STM) overview map showing that bismuthene fully covers the substrate. The flakes are of ~25-nm extent, limited by domain boundaries. (c) Substrate step-height profile, taken along the dark line in (b). The step heights correspond to SiC steps. (d) The honeycomb pattern is seen on smaller scan frames. (e) Close-up STM images for occupied and empty states (left and right panels, respectively). They confirm the formation of Bi honeycombs.[20]

quasi-2D kagome-derived topological bands with charge-density-wave order, nematicity, and superconductivity.[28,29]

4.2.3 Low-Dimensional Topological Semiconductors

Low-dimensional topological semiconductors occupy an important intermediate position because they combine the tunability of conventional semiconductor platforms with nontrivial band topology. In many cases, their topological properties arise from controllable band inversion induced by confinement, strain, layer thickness, or electrostatic gating. The HgTe/CdTe and InAs/GaSb quantum-well systems already demonstrated that semiconductor heterostructures can be engineered into topological regimes through precise control of band alignment and structure.[1,18]

This tunability is especially important from the perspective of applications because semiconductor-based platforms are naturally compatible with established device architectures. More recently, the concept of topological semiconductors has expanded into van der Waals heterostructures and moiré superlattices, where the electronic structure can be continuously reshaped by twist angle, displacement field, and interaction strength.[30] In such systems, topology is no longer a fixed material property, but a tunable feature of an engineered band structure.

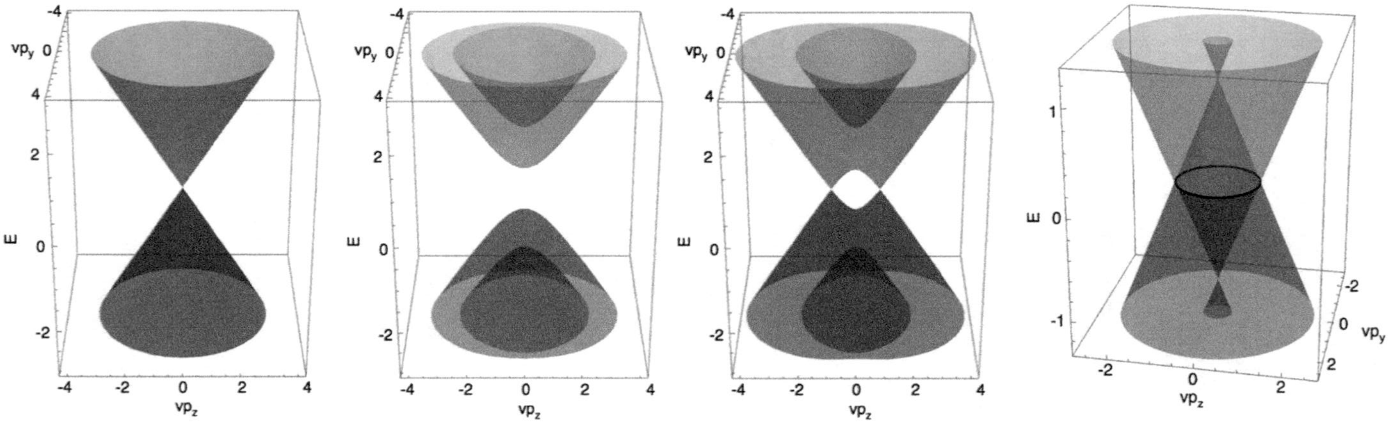

FIGURE 4.9 (Left to right) Energy spectra of $\varepsilon_{s\mu}$ $(0,\ p_y,\ p_z)$ for the Dirac semimetal ($m=b=b'=0$), magnetic semiconductor ($m=1$, $b=0.5$, $b'=0$), Weyl semimetal ($m=0.5$, $b=1$, $b'=0$), and line node semimetal ($m=0$, $b=0$, $b'=1$) for the Hamiltonian equation.[24]

A representative recent example is twisted bilayer InSe, in which theory identified a time-reversal-invariant topological state and corresponding edge states in a 2D semiconductor moiré system.[31] This result is important because it extends the idea of low-dimensional topological semiconductors from conventional quantum wells to artificial moiré semiconductor platforms.

4.2.4 Low-Dimensional Topological Superconductors

Low-dimensional topological superconductors host superconducting states with non-trivial topology and are predicted to support Majorana bound states, which obey non-Abelian statistics and are of central interest for fault-tolerant quantum computation.[13] The conceptual starting point of this field is the Kitaev chain, which showed that a 1D p-wave superconducting wire can host zero-energy Majorana modes localized at its ends.[32] A more experimentally accessible route was later proposed by Fu and Kane, who showed that Majorana modes can emerge at the interface between an ordinary superconductor and a topological insulator.[33]

The first widely discussed experimental signatures of Majorana zero modes were reported in hybrid semiconductor–superconductor nanowire devices, where zero-bias conductance anomalies consistent with Majorana physics were observed.[34] Subsequently, iron-based superconductors emerged as an important alternative platform because they may intrinsically combine superconductivity and topological band structure within a single material. In particular, Fe(Se, Te) has attracted major attention because scanning tunneling spectroscopy revealed zero-energy vortex-core states consistent with Majorana bound states.[35,36]

Recent studies have further strengthened the relevance of $FeSe_{1-x}$Tex and related systems by emphasizing the interplay between topology, superconductivity, and broken symmetries.[37] At the same time, growing attention has been paid to correlation-tuned topological superconductivity, where interaction effects may actively stabilize nontrivial superconducting states.[38]

4.2.5 Low-Dimensional Topological Magnetic Materials

When nontrivial band topology is combined with magnetic order, time-reversal symmetry is broken, and new topological phases become possible, including Chern insulators and the quantum anomalous Hall effect. The quantum anomalous Hall effect was first experimentally realized in magnetically doped topological-insulator thin films, where dissipationless chiral edge transport was observed without an external magnetic field.[14]

In recent years, extensive work has focused on improving the materials quality and thin-film growth of $MnBi_2Te_4$, especially in the context of realizing more robust quantum anomalous Hall behavior.[40–42] At the same time, an important new direction has emerged in the form of altermagnetism. Altermagnets possess zero net magnetization but exhibit spin-split electronic bands, making them conceptually distinct from both ferromagnets and conventional antiferromagnets.[43,44]

A major subsequent advance was the emergence of intrinsic magnetic topological materials, in which magnetism is built into the crystal itself rather than introduced through chemical doping. Among these, $MnBi_2Te_4$ has become the flagship system. It was identified as an intrinsic magnetic topological insulator and soon shown to host thickness-dependent quantum anomalous Hall states and related magnetic topological phases in Figure 4.10.[39] As it is a layered van der Waals compound, $MnBi_2Te_4$ also provides a particularly appealing low-dimensional platform for investigating the interplay between magnetism, topology, and thickness control.

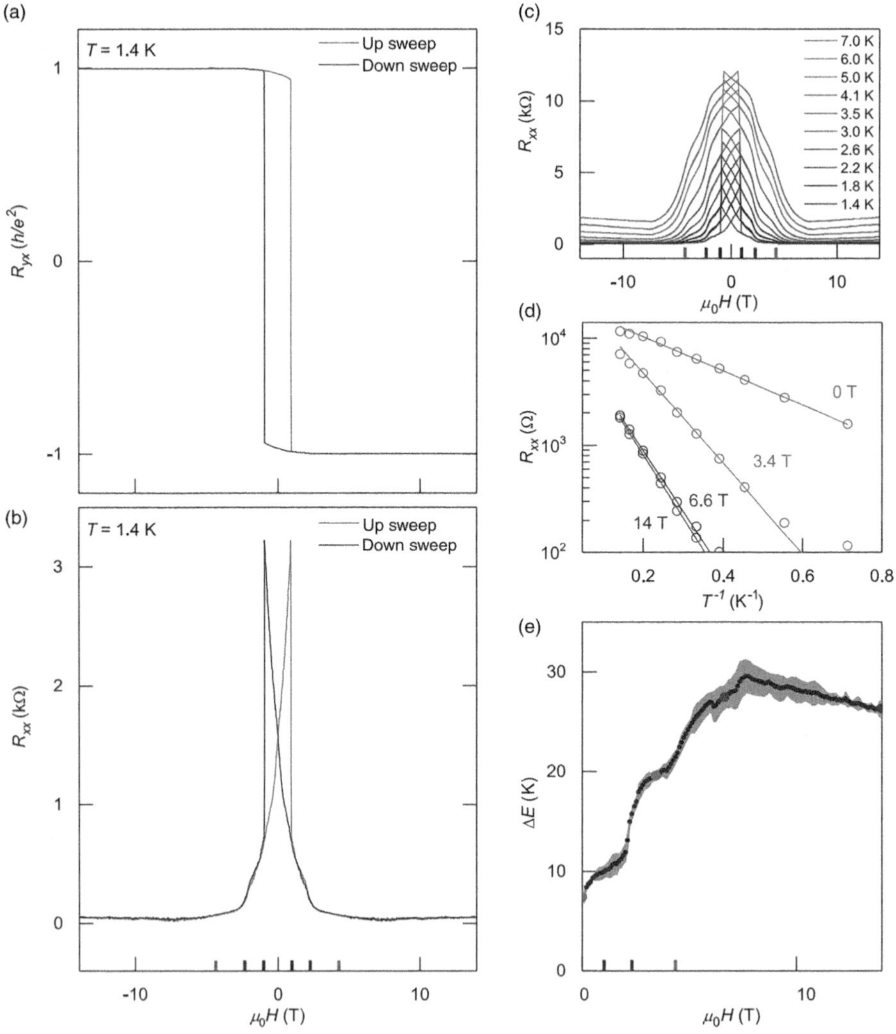

FIGURE 4.10 QAH effect in a five-layer $MnBi_2Te_4$ flake.[39]

4.2.6 Low-Dimensional Topological Photonic Materials

Topological ideas have also been extended beyond fermionic condensed-matter systems to bosonic platforms, especially photonic crystals, waveguide arrays, and microwave lattices. In topological photonic materials, synthetic band structures are engineered to emulate electronic topological phases, while protected edge transport can be observed directly through light propagation.[45] As coupling geometry, symmetry, and dimensionality can be precisely controlled, photonic systems have become an exceptionally clean setting in which to test topological principles.

Early landmark experiments demonstrated topological edge transport in silicon photonics and photonic Floquet lattices, establishing that robust boundary propagation can occur even in the absence of conventional electronic degrees of freedom.[46,47] Since then, topological photonics has expanded rapidly toward nonlinear, non-Hermitian, and higher-order regimes. Recent work has increasingly emphasized reconfigurable and higher-order topological photonic structures. Synthetic-dimension and polycrystal-based photonic systems have provided new routes toward tunable corner, edge, and interface states in integrated optical architectures.[48]

4.2.7 Other Low-Dimensional Topological Materials

Beyond the major categories discussed above, a growing number of low-dimensional systems exhibit topological properties that do not fit neatly into the conventional classifications of topological insulators, semimetals, superconductors, magnetic topological materials, or photonic topological systems. These include higher-order topological materials, moiré topological systems, topological phononic and magnonic structures, and artificially engineered lattices and metamaterials. Although the microscopic degrees of freedom in these systems may differ substantially from those in electronic materials, they share a common topological origin: nontrivial band structures and protected boundary excitations governed by symmetry and geometry.[49,50]

One important recent direction is the study of higher-order topological phases, in which protected states do not appear on ordinary edges or surfaces but instead emerge at higher codimension boundaries, such as corners or hinges. This concept extends the conventional bulk–boundary correspondence and has significantly broadened the understanding of topology in reduced-dimensional systems. The theoretical framework of higher-order topology was established by the quadrupole-insulator model and related developments.[49,50] In low-dimensional settings, higher-order topology has subsequently been explored in electronic materials, photonic crystals, acoustic metamaterials, and circuit-based platforms, where corner states can often be detected more directly than in bulk solids.[51–53]

Another rapidly developing area is moiré topological matter, in which topology emerges from twist angle, interlayer registry, and interaction-driven band reconstruction in stacked 2D materials. Moiré superlattices can generate flat bands, enhanced correlations, and tunable Berry curvature, making them an unusually flexible platform for realizing topological phases. Recent work has established that moiré graphene and related twisted van der Waals systems can host Chern insulating states and other interaction-driven topological phases.[29,30,54] While some of these systems may also be grouped under topological semiconductors or correlated topological materials, it is often useful to regard them more broadly as part of a wider class of engineered low-dimensional topological platforms.

In addition, topology has been extended to phononic, acoustic, and magnonic systems, where protected boundary transport can occur for collective excitations rather than electrons. These systems are particularly valuable because they provide highly controllable realizations of topological concepts and often allow direct visualization of boundary modes. Representative examples include topological acoustic metamaterials and magnonic systems with symmetry-protected boundary transport.[51,52,55] Artificially designed lattices and metamaterials also play an increasingly important role in this context, as they permit independent control over geometry, coupling, and symmetry, thereby enabling systematic exploration of otherwise difficult-to-access topological regimes.[51–53]

Taken together, these systems demonstrate that low-dimensional topology is no longer limited to conventional electronic materials. Instead, it has evolved into a broader framework that encompasses both natural and engineered quantum platforms. This expansion is important for two reasons: first, it reveals the universality of topological band theory across very different physical systems; second, it provides new opportunities for designing tunable topological functionalities beyond the constraints of traditional crystals.[49–55]

4.3 REGULATION OF LOW-DIMENSIONAL TOPOLOGICAL MATERIALS

A defining feature of low-dimensional topological materials is that their topological phases are often highly tunable. Because the relevant band inversion, symmetry protection, Berry-curvature distribution, and boundary-state transport all occur on energy scales accessible to external perturbations, low-dimensional systems provide a particularly favorable platform for regulating topology via magnetic fields, electric fields, strain, chemical doping, and other means. In contrast to bulk materials, where topological phases are often largely fixed by the intrinsic crystal structure, low-dimensional materials can frequently be driven across topological phase boundaries by relatively modest external stimuli. This tunability is important not only for clarifying the microscopic origin of topological phenomena but also for enabling switchable topological devices and multifunctional quantum architectures.

4.3.1 Magnetic Field Modulation

Magnetic-field modulation is one of the most direct approaches for controlling topology because magnetic fields can break time-reversal symmetry, reconstruct Landau levels, modify spin polarization, and drive transitions among quantum spin Hall, Chern-insulating, axion-insulating, and trivial regimes. In low-dimensional topological systems, these effects are especially pronounced because orbital quantization, Zeeman splitting, and edge-state transport all strongly depend on reduced dimensionality.

A representative example is provided by HgTe-based quantum wells. Beyond their foundational role in quantum spin Hall physics, magnetically modified HgTe systems have shown that quantized spin Hall transport can persist even in the presence of dilute magnetic Mn dopants, indicating that magnetic perturbations can be used not only to suppress but also to reshape topological transport in a controllable manner.[42] In hybrid topological Josephson structures based on dilute magnetic semiconducting weak links, the magnetic field has further been used to tune supercurrent behavior, showing that the combination of topology, magnetism, and superconducting proximity can generate field-responsive quantum transport functionalities.[57]

An even more important recent platform for magnetic-field regulation is $MnBi_2Te_4$, the prototypical intrinsic magnetic topological insulator. In this material family, the magnetic field can tune the balance between antiferromagnetic and spin-polarized states, thereby driving transitions among axion-insulator, zero-Hall-plateau, and Chern-insulator regimes. High-field transport studies have shown that few-layer $MnBi_2Te_4$ exhibits robust field-induced topological phase evolution, confirming that the magnetic field acts as a powerful knob for controlling the Chern number and the associated chiral transport channels, as shown in Figure 4.11.[56,58,59]

4.3.2 Electric Field Modulation

Electric fields provide perhaps the most technologically attractive route for regulating low-dimensional topological materials because they allow in situ, reversible, and low-power control through electrostatic gating or displacement fields. In essence, electric fields modify topology by shifting orbital energies, breaking inversion symmetry, redistributing charge, and tuning the relative position of topological and trivial bands. In low-dimensional materials, where screening is weak and electrostatic coupling is strong, these effects can be substantial even at moderate gate voltages.

Monolayer 1T′-WTe_2 offers a representative example. Scanning tunneling spectroscopy studies have shown that gate-induced electric fields can strongly modify its bulk gap, demonstrating directly that electrostatic control can regulate the underlying band structure of a 2D QSHI as shown in Figure 4.12.[50] Similarly, in a 2D topological insulator heterostructure based on cadmium arsenide, a tunable displacement field was shown to convert a crossing of zeroth Landau levels into an avoided crossing, signaling a change from inverted to trivial band order.[60] These results are particularly important because they show that electric fields do not merely perturb transport but can drive actual topological phase transitions.

More broadly, recent roadmaps for 2D topological insulators identify electrically controlled switching and topological phase engineering as major directions for the future device development.[61] In van der Waals and moiré semiconductor systems, gate-controlled displacement fields and twist-dependent electrostatics now make it possible to approach electrically programmable topological phases in increasingly realistic device architectures.

4.3.3 Strain Modulation

Strain modulation is especially powerful in low-dimensional materials because their reduced thickness and mechanical flexibility make them highly responsive to lattice deformation. Strain can modify bond lengths, crystal-field splitting, SOC strength,

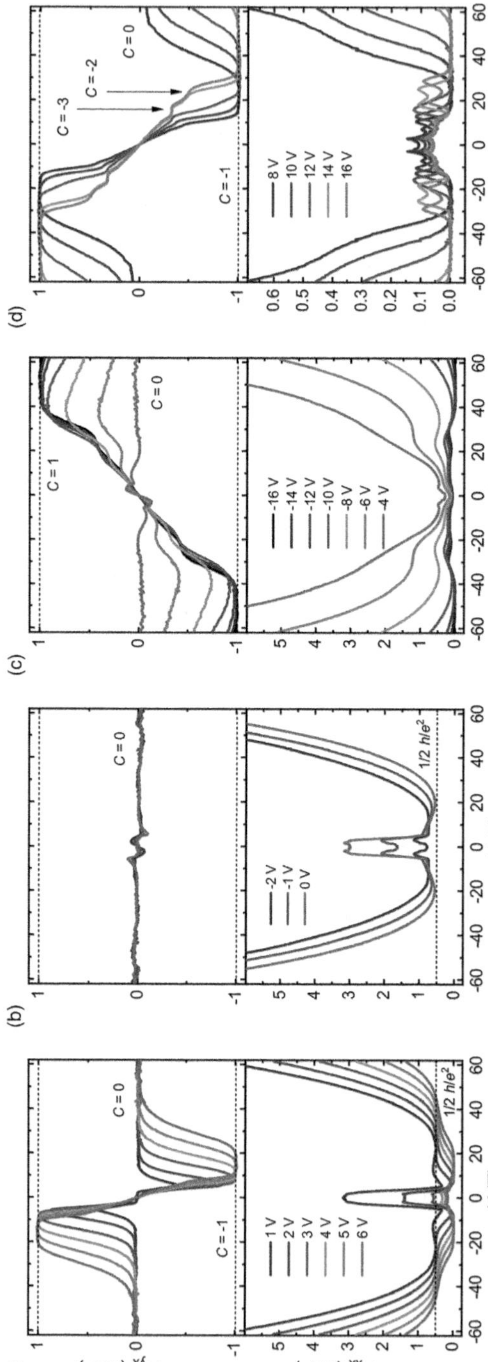

FIGURE 4.11 Transport properties in pulsed magnetic field up to 61.5 T.[56]

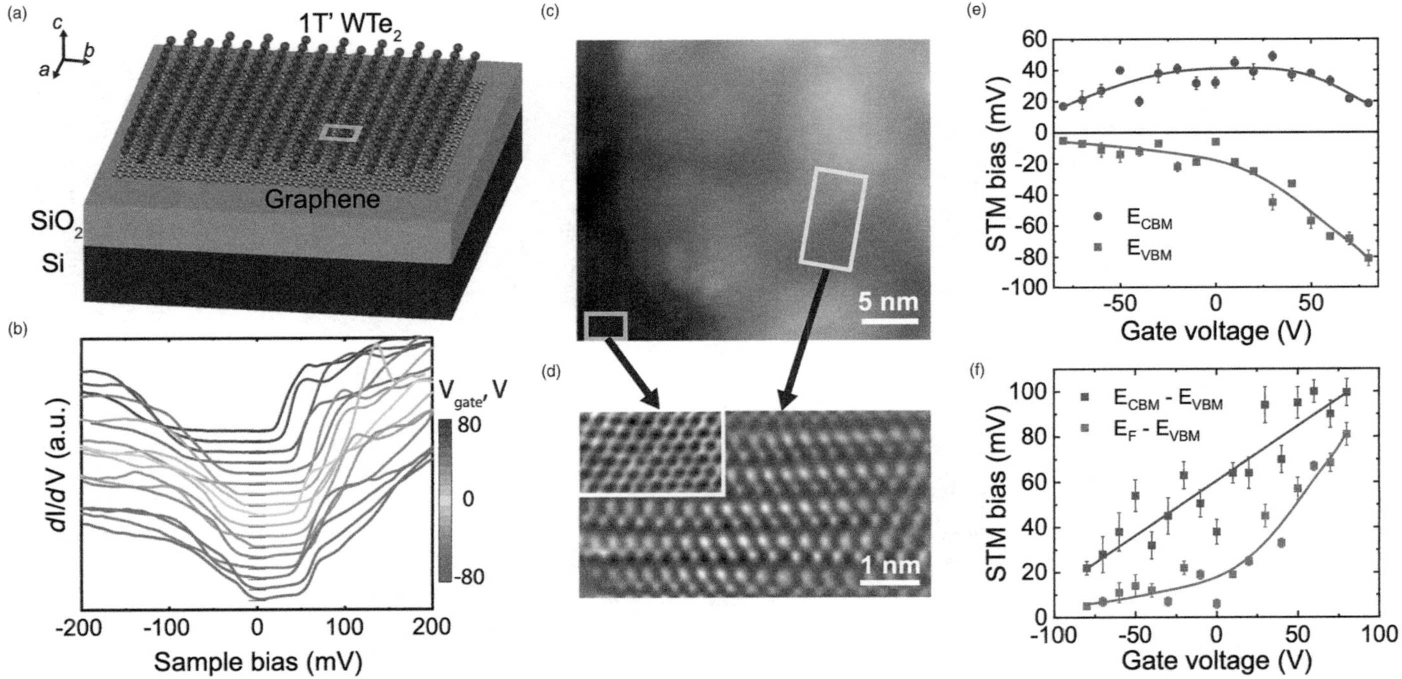

FIGURE 4.12 Back-gated STM measurements of ML-WTe_2 on graphene on SiO_2. (a) Schematic of the back-gating setup. (b) Evolution of the bulk dI / dV spectrum of a monolayer island in response to the gate voltage. (c) STM image of monolayer WTe_2 island on graphene on SiO_2. (d) A zoom-in with atomic resolution showing Te atoms as well as the graphene lattice (inset). (e) Plots of the conduction band minimum and the valence band maximum for each spectrum at the given gate voltage (dots corresponding to (ECBM) and (EVBM), respectively) showing the nontrivial evolution of the bulk electronic structure with gate voltage. The error bars are determined by the energy intervals at which data were obtained and the noise level of the dI / dV spectra. (f) Same as (e) but with energies relative to the VBM. Back-gating changes the band gap (dots representing (ECBM-EVBM)) and the position of the VBM with respect to E_F (dots representing (EF-EVBM)).[50]

and lattice symmetry, and thus directly affect band inversion and Berry-curvature distribution. In many cases, strain can drive a transition between trivial and topological phases without introducing chemical disorder.

Recent reviews of strain-tunable quantum phenomena in low-dimensional materials emphasize that strain has become one of the most versatile tools for engineering topological transitions and topological functionalities.[52] In the context of low-dimensional topological materials, recent work on bismuthene has shown that the combination of strain and periodic optical driving can manipulate topological edge states and induce Floquet topological phase transitions (Figure 4.13).[53] Although many strain-driven phase transitions are still explored theoretically, the physical mechanism is now well established: strain shifts the relative energies of the relevant orbitals, changes band inversion thresholds, and can stabilize or suppress protected boundary modes.

Strain is also attractive from an application perspective because it offers a continuous and, in principle, spatially programmable form of control. Unlike chemical substitution, strain can be applied locally and reversibly, making it promising for flexible

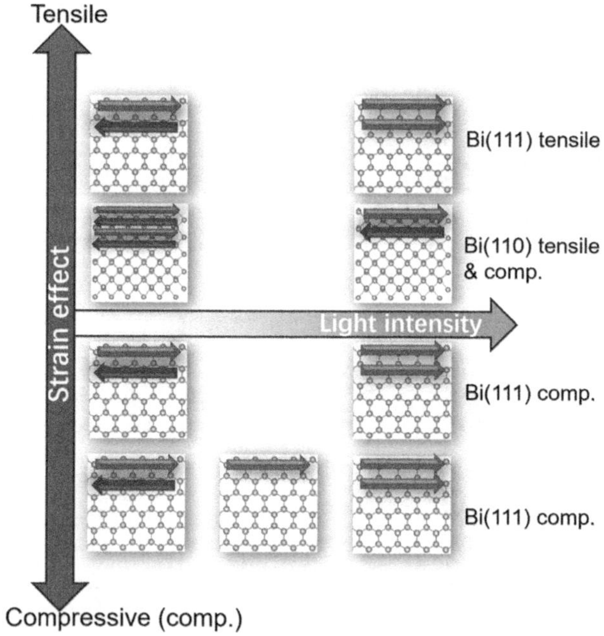

FIGURE 4.13 By considering the combination of strain effects and external periodic light fields, the nontrivial edge states in nanoribbons of bismuthene are tunable. For the Bi(110) monolayer, double pairs of gapless helical edge states will evolve into a pair of helical edge states with increasing light intensity. For the Bi(111) monolayer, the topologically protected helical edge states will evolve into two chiral edge states with a Chern number $C = 2$; however, with a larger compressive (comp.) strain effect and increasing light intensity, the helical edge states could be manipulated into one and two chiral edge states, successively.[53]

electronics, strain-sensitive topological sensors, and heterostructures with spatially patterned topological phases.[52]

4.3.4 Doping Modulation

Doping modulation remains one of the most widely used strategies for regulating low-dimensional topological materials because it can alter carrier density, exchange interaction, SOC, and local symmetry. Depending on dopant type and concentration, doping may preserve the parent topology, shift the Fermi level into the regime of protected states, induce magnetic order, or even trigger a topological phase transition.

A classic example is magnetic doping in topological-insulator thin films, which enabled the first realization of the quantum anomalous Hall effect by introducing ferromagnetism into an originally time-reversal-invariant topological system.[14] More recently, doping has also been discussed in a broader topological-phase-engineering sense. For instance, carrier doping has been proposed as a route to transform Bi from a higher-order topological-insulator phase into a more conventional topological regime, while magnetic doping of dual-topological systems such as Bi_2TeI has been predicted to induce transitions toward magnetic Weyl phases.[53,57]

In practical terms, doping remains valuable because it is compatible with standard synthesis and thin-film growth techniques. However, it also introduces a key trade-off: the same dopants that enable desirable band or magnetic engineering may simultaneously enhance disorder and degrade mobility. For this reason, contemporary strategies increasingly emphasize modulation doping, interface doping, or controlled dilute substitution, especially in 2D materials and heterostructures where preserving high-quality edge or surface transport is essential.[59]

4.3.5 Other Regulation Methods

Beyond magnetic field, electric field, strain, and doping, a growing number of additional regulation methods are being explored in low-dimensional topological materials. Among these, pressure, twist-angle engineering, light-driven Floquet control, proximity coupling, and heterostructure design are especially important. These methods regulate topology through more indirect but highly effective routes: pressure changes lattice parameters and bandwidth, twist engineering reshapes moiré band topology, light periodically dresses electronic states, and proximity effects transfer exchange, superconducting, or ferroelectric order across interfaces.

A representative recent example is twisted bilayer InSe, where theory provided evidence for a time-reversal-invariant topological state and corresponding edge states in a 2D semiconductor moiré platform (Figure 4.14).[31] This result is important because it demonstrates that topology can emerge not only from the intrinsic chemistry of a single layer but also from the geometry of interlayer twist. Likewise, recent photonic and Floquet studies show that time-periodic driving can be used to induce or manipulate topological edge states, further expanding the concept of regulation from static materials engineering to dynamic topological control.[48,53]

Taken together, these developments indicate that the regulation of low-dimensional topological materials is no longer limited to a few conventional external fields.

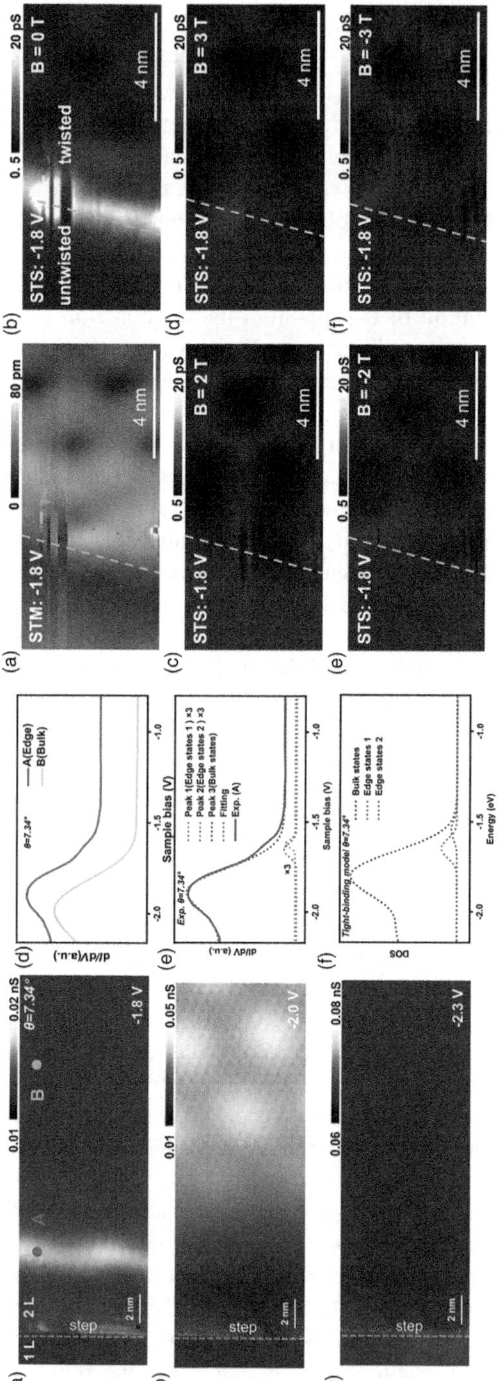

FIGURE 4.14 Edge states of bilayer InSe with 7.34° twisting angle and topological nature of the twisted bilayer InSe.[31]

Instead, the field is moving toward multimodal regulation, in which several perturbations (such as gating, twist, strain, light, and magnetic proximity) are combined to tailor topology on demand. This trend is likely to be crucial for the eventual realization of topological transistors, programmable topological circuits, and quantum-information devices.[48,61]

4.4 APPLICATION OF LOW-DIMENSIONAL TOPOLOGICAL MATERIALS

The application prospects of low-dimensional topological materials arise directly from the same features that make them scientifically distinctive: protected boundary conduction, spin-momentum locking, large Berry-curvature effects, enhanced surface or edge contributions, and strong compatibility with heterostructure engineering. In low-dimensional form, these advantages are often amplified because the fraction of carriers associated with edges, interfaces, or surface states becomes much larger than in bulk materials. As a result, many proposed applications are not merely extensions of bulk topological physics but are fundamentally enabled by reduced dimensionality.

One of the most intensively discussed directions is low-power electronics and spintronics. The 2D topological insulators are attractive because their helical edge states offer dissipationless or low-dissipation transport channels protected against backscattering by time-reversal symmetry. The 2024 roadmap on 2D topological insulators explicitly highlights their promise for electronic, spintronic, and quantum-device applications, especially when combined with ferromagnets, multiferroics, or superconductors in heterostructures.[49] Magnetic topological materials extend this idea by enabling chiral edge transport and the quantum anomalous Hall effect without an external field, which is highly relevant for ultralow-power interconnects and nonvolatile spintronic functionalities.

A second major direction is topological quantum computation. Low-dimensional topological superconductors and topological-insulator/superconductor hybrid structures are widely pursued because they can host Majorana zero modes, which are promising building blocks for fault-tolerant qubits. Here, the application value depends critically on controllability: electric, magnetic, and proximity regulation are required not only to create Majorana-supporting phases but also to manipulate them in device geometries. For this reason, the boundary between regulation and application is especially blurred in the quantum-computing context: the same tuning knobs that reveal topological superconductivity are also the basis of the future quantum functionality.

Low-dimensional topological materials also show considerable promise in optoelectronics and photonics (Figure 4.15).[48] Topological photonic systems offer robust light propagation protected against certain classes of disorder and backscattering, making them appealing for integrated optical routing, reconfigurable photonic circuits, and topological lasers. In parallel, reviews of nanostructured topological materials emphasize their relevance to optoelectronic devices because low-dimensional

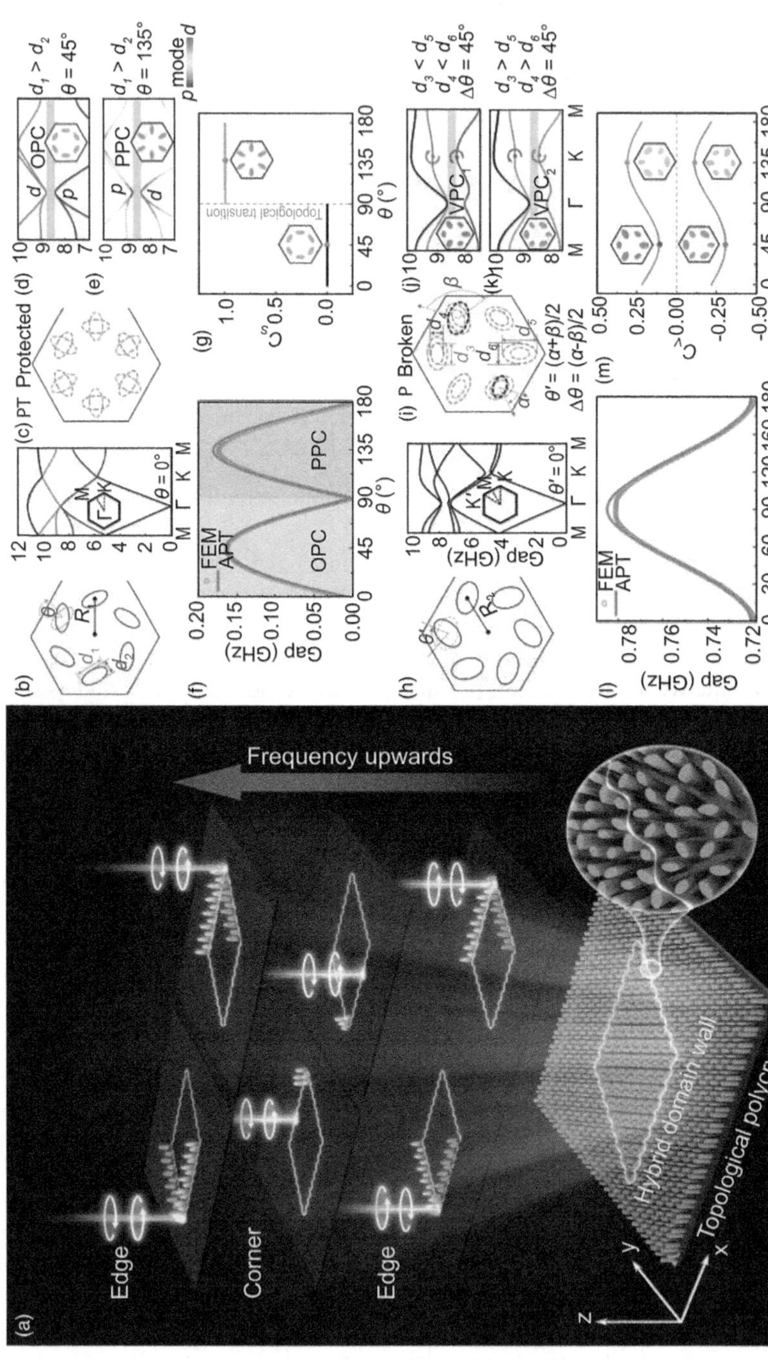

FIGURE 4.15 The topological photonic polycrystal (TPPC) applied to photonic integrated circuitry (PIC) and its detailed composition.[48]

geometries facilitate integration and enhance the relative role of topological boundary channels.[61]

Another important application area is thermoelectrics and magneto-thermoelectrics. Topological insulators and semimetals possess unusual transport features, such as high mobility, band inversion, and strong Berry-curvature responses, that can enhance Seebeck, Nernst, or related thermoelectric effects, as shown in Figure 4.16.[62] Recent reviews on topological thermoelectrics emphasize that both topological insulators and topological semimetals offer new routes to enhance thermoelectric performance, especially when low dimensionality and nanostructuring are used to optimize phonon scattering and electronic transport simultaneously.[63]

Finally, low-dimensional topological materials are increasingly relevant for multifunctional heterostructures and emerging quantum devices. The combination of topology with ferromagnetism, antiferromagnetism, superconductivity, ferroelectricity, or moiré engineering enables composite functionalities that are difficult to achieve in any single conventional material platform.[59,61] This is likely to be one of

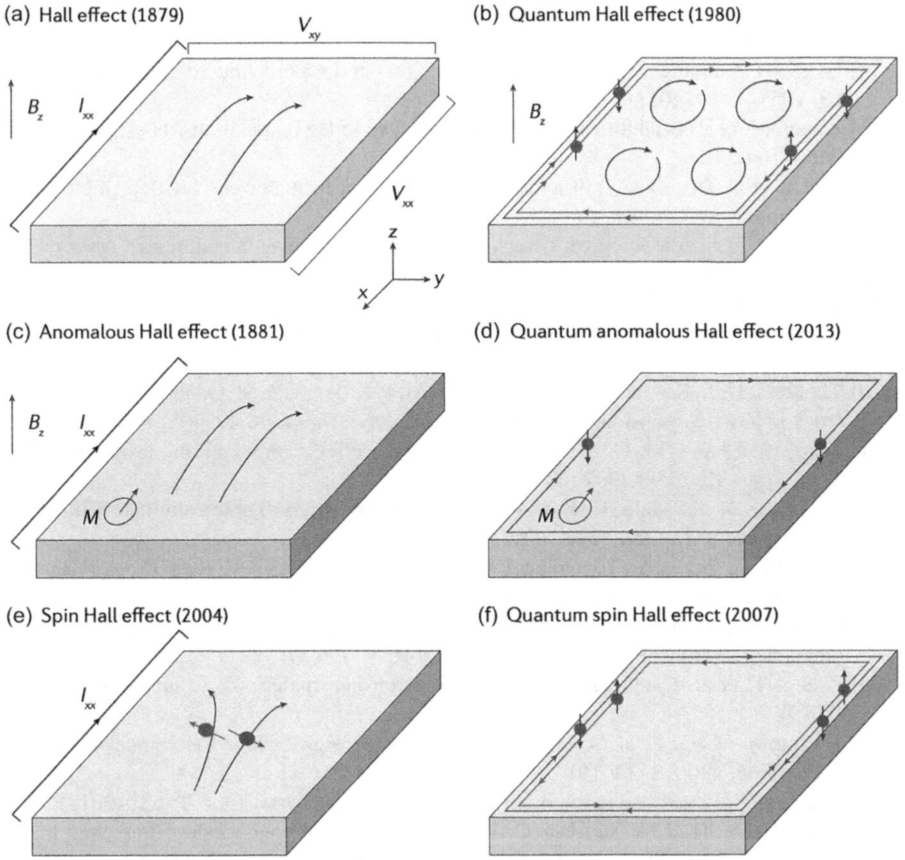

FIGURE 4.16 Different classes of 2D topological insulators, and their classical counter parts.[62]

the most important long-term application routes: rather than serving only as isolated topological conductors, low-dimensional topological materials may function as active layers in complex, tunable, and multifunctional quantum architectures. Overall, their practical significance lies not only in their protected states but in the fact that these states can be engineered, regulated, and integrated into device-relevant geometries. Continued progress in materials synthesis, disorder control, interface design, and external-field tuning will therefore determine how rapidly the field moves from proof-of-principle demonstrations toward scalable topological technologies.

REFERENCES

[1] Bernevig, B. A., Hughes, T. L. & Zhang, S.-C. Quantum spin Hall effect and topological phase transition in HgTe quantum wells. *Science* 314, 1757–1761 (2006).
[2] Kane, C. L. & Mele, E. J. Quantum spin Hall effect in graphene. *Phys. Rev. Lett.* 95, 226801 (2005).
[3] Kane, C. L. Topological order and the quantum spin Hall effect. *Phys. Rev. Lett.* 95, 146802 (2005).
[4] Hasan, M. Z. & Kane, C. L. Colloquium: Topological insulators. *Rev. Mod. Phys.* 82, 3045–3067 (2010).
[5] Qi, X.-L. & Zhang, S.-C. Topological insulators and superconductors. *Rev. Mod. Phys.* 83, 1057–1110 (2011).
[6] König, M. et al. Quantum spin hall insulator state in HgTe quantum wells. *Science* 318, 766–770 (2007).
[7] Xia, Y. et al. Observation of a large-gap topological-insulator class with a single Dirac cone on the surface. *Nat. Phys.* 5, 398–402 (2009).
[8] Hsieh, D. et al. A topological Dirac insulator in a quantum spin Hall phase. *Nature* 452, 970–974 (2008).
[9] Klitzing, K. V., Dorda, G. & Pepper, M. New method for high-accuracy determination of the fine-structure constant based on quantized hall resistance. *Phys. Rev. Lett.* 45, 494–497 (1980).
[10] Thouless, D. J., Kohmoto, M., Nightingale, M. P. & den Nijs, M. Quantized hall conductance in a two-dimensional periodic potential. *Phys. Rev. Lett.* 49, 405–408 (1982).
[11] Xiao, D., Chang, M.-C. & Niu, Q. Berry phase effects on electronic properties. *Rev. Mod. Phys.* 82, 1959–2007 (2010).
[12] Armitage, N. P., Mele, E. J. & Vishwanath, A. Weyl and Dirac semimetals in three-dimensional solids. *Rev. Mod. Phys.* 90, 015001 (2018).
[13] Sato, M. & Ando, Y. Topological superconductors: A review. *Rep. Prog. Phys.* 80, 076501 (2017).
[14] Chang, C.-Z. et al. Experimental observation of the quantum anomalous hall effect in a magnetic topological insulator. *Science* 340, 167–170 (2013).
[15] Zhang, T. et al. Catalogue of topological electronic materials. *Nature* 566, 475–479 (2019).
[16] Vergniory, M. G. et al. A complete catalogue of high-quality topological materials. *Nature* 566, 480–485 (2019).
[17] Fei, Z. et al. Edge conduction in monolayer WTe_2. *Nat. Phys.* 13, 677–682 (2017).
[18] Knez, I., Du, R.-R. & Sullivan, G. Evidence for helical edge modes in inverted InAs/GaSb quantum wells. *Phys. Rev. Lett.* 107, 136603 (2011).
[19] Tang, S. et al. Quantum spin Hall state in monolayer 1T'-WTe_2. *Nat. Phys.* 13, 683–687 (2017).

[20] Reis, F. et al. Bismuthene on a SiC substrate: A candidate for a high-temperature quantum spin Hall material. *Science* 357, 287–290 (2017).

[21] Gehrig, L. et al. Bismuthene under cover: Graphene intercalation of a large gap quantum spin hall insulator. *Adv. Mater.* 37, 2502412 (2025).

[22] Meyer, M. et al. Quantum spin Hall effect in III-V semiconductors at elevated temperatures: Advancing topological electronics. *Sci. Adv.* 11, eadz2408 (2025).

[23] Liu, Z. K. et al. Discovery of a three-dimensional topological Dirac semimetal, Na_3Bi. *Science* 343, 864–867 (2014).

[24] Koshino, M. & Hizbullah, I. F. Magnetic susceptibility in three-dimensional nodal semimetals. *Phys. Rev. B* 93, 045201 (2016).

[25] Borisenko, S. et al. Experimental realization of a three-dimensional Dirac semimetal. *Phys. Rev. Lett.* 113, 027603 (2014).

[26] Xu, S.-Y. et al. Discovery of a Weyl fermion semimetal and topological Fermi arcs. *Science* 349, 613–617 (2015).

[27] Lv, B. Q. et al. Experimental discovery of Weyl semimetal TaAs. *Phys. Rev. X* 5, 031013 (2015).

[28] Jiang, K. et al. Kagome superconductors AV_3Sb_5 (A = K, Rb, Cs). *Natl. Sci. Rev.* 10, nwac199 (2023).

[29] Xu, Z., Le, T. & Lin, X. AV_3Sb_5 Kagome superconductors: A review with transport measurements. *Chin. Phys. Lett.* 42, 037304 (2025).

[30] Waters, D. et al. Chern insulators at integer and fractional filling in moiré pentalayer graphene. *Phys. Rev. X* 15, 011045 (2025).

[31] Tian, D. et al. Real-space observation of a time-reversal invariant topological state in twisted bilayer InSe. *Nat. Commun.* 16, 9453 (2025).

[32] Kitaev, A. Y. Unpaired Majorana fermions in quantum wires. *Phys.-Usp.* 44, 131 (2001).

[33] Fu, L. & Kane, C. L. Superconducting proximity effect and Majorana fermions at the surface of a topological insulator. *Phys. Rev. Lett.* 100, 096407 (2008).

[34] Mourik, V. et al. Signatures of Majorana fermions in hybrid superconductor-semiconductor nanowire devices. *Science* 336, 1003–1007 (2012).

[35] Wang, D. et al. Evidence for Majorana bound states in an iron-based superconductor. *Science* 362, 333–335 (2018).

[36] Machida, T. et al. Zero-energy vortex bound state in the superconducting topological surface state of Fe(Se,Te). *Nat. Mater.* 18, 811–815 (2019).

[37] Roppongi, M. et al. Topology meets time-reversal symmetry breaking in $FeSe_{1-x}Te_x$ superconductors. *Nat. Commun.* 16, 6573 (2025).

[38] Lin, H. et al. A topological superconductor tuned by electronic correlations. *Nat. Commun.* 17, 1188 (2025).

[39] Deng, Y. et al. Quantum anomalous Hall effect in intrinsic magnetic topological insulator MnBi2Te4. *Science* 367, 895–900 (2020).

[40] Li, S. et al. Progress on the antiferromagnetic topological insulator $MnBi_2Te_4$. *Natl. Sci. Rev.* 11, nwac296 (2024).

[41] Li, Q., Mo, S.-K. & Edmonds, M. T. Recent progress of $MnBi_2Te_4$ epitaxial thin films as a platform for realising the quantum anomalous Hall effect. *Nanoscale* 16, 14247–14260 (2024).

[42] Vyazovskaya, A. Y., Bosnar, M., Chulkov, E. V. & Otrokov, M. M. Intrinsic magnetic topological insulators of the $MnBi_2Te_4$ family. *Commun. Mater.* 6, 88 (2025).

[43] Cheong, S.-W. & Huang, F.-T. Altermagnetism classification. *NPJ Quantum Mater.* 10, 38 (2025).

[44] Song, C. et al. Altermagnets as a new class of functional materials. *Nat. Rev. Mater.* 10, 473–485 (2025).

[45] Lu, L., Joannopoulos, J. D. & Soljačić, M. Topological photonics. *Nat. Photon.* 8, 821–829 (2014).
[46] Hafezi, M., Mittal, S., Fan, J., Migdall, A. & Taylor, J. M. Imaging topological edge states in silicon photonics. *Nat. Photon.* 7, 1001–1005 (2013).
[47] Rechtsman, M. C. et al. Photonic Floquet topological insulators. *Nature* 496, 196–200 (2013).
[48] Li, T. et al. Configurable topological photonic polycrystal based on a synthetic hybrid dimension. *Natl. Sci. Rev.* 12, nwaf107 (2025).
[49] Weber, B. et al. 2024 roadmap on 2D topological insulators. *J. Phys. Mater.* 7, 022501 (2024).
[50] Maximenko, Y. et al. Nanoscale studies of electric field effects on monolayer 1T′-WTe_2. *NPJ Quantum Mater.* 7, 29 (2022).
[51] Rashidi, A., Ahadi, S., Munyan, S., Mitchell, W. J. & Stemmer, S. Tuning displacement fields in a two-dimensional topological insulator using nanopatterned gates. *Nano Lett.* 24, 7366–7372 (2024).
[52] Kim, J. M. et al. Strain engineering of low-dimensional materials for emerging quantum phenomena and functionalities. *Adv. Mater.* 35, 2107362 (2023).
[53] Kong, X., Zhang, B., Gong, W. & Li, L. Tunable light-induced topological edge states in strain engineering of bismuthene monolayers. *J. Mater. Chem. C* 12, 13325–13331 (2024).
[54] Shamim, S. et al. Quantized spin Hall conductance in a magnetically doped two dimensional topological insulator. *Nat. Commun.* 12, 3193 (2021).
[55] Mandal, P. et al. Magnetically tunable supercurrent in dilute magnetic topological insulator-based Josephson junctions. *Nat. Phys.* 20, 984–990 (2024).
[56] Liu, C. et al. Magnetic-field-induced robust zero Hall plateau state in $MnBi_2Te_4$ Chern insulator. *Nat. Commun.* 12, 4647 (2021).
[57] Jin, K.-H. Doping-induced topological phase transition in Bi: The role of quantum electronic stress. *Phys. Rev. B* 101, 035111 (2020).
[58] Bhattarai, D., Shahi, D. B., Lawaju, S., Mitra, C., & Ghimire, M. P. Magnetic doping-induced topological phase transition in Bi_2TeI: A density functional study. *The Journal of Physical Chemistry C* 129, 8764–8770 (2025).
[59] Wang, D., Li, X.-B. & Sun, H.-B. Modulation doping: A strategy for 2D materials electronics. *Nano Lett.* 21, 6298–6303 (2021).
[60] Jiang, Q., Chatterjee, A. & Huang, S. Synthesis and characterization of nanostructured topological materials. *Prog. Quantum Electron.* 103, 100588 (2025).
[61] Toriyama, M. Y. & Snyder, G. J. Topological insulators for thermoelectrics: A perspective from beneath the surface. *Innovation* 6, 100782 (2025).
[62] von Klitzing, K. et al. 40 years of the quantum Hall effect. *Nat. Rev. Phys.* 2, 397–401 (2020).
[63] Yang, T., Yang, Y., Wang, X., Zhang, G. & Cheng, Z. Topological thermoelectrics: New opportunities and challenges. *Mater. Today Chem.* 30, 101488 (2023).

5 Low-Dimensional Superconducting Materials

Yi Zhang and Long Ren

5.1 BASIC OVERVIEW

Superconductivity in reduced-dimensional systems has attracted considerable attention in contemporary quantum materials research. In these systems, superconductivity can emerge in one- or few-atomic-layer structures, interfacial heterostructures, or quasi-one-dimensional (quasi-1D) geometries. Compared with their bulk counterparts, such reduced-dimensional superconductors often exhibit modified physical properties arising from confinement effects, enhanced quantum fluctuations, reduced electronic screening, and interfacial coupling. In addition, spin–orbit interaction and symmetry breaking in low dimensions can further influence the superconducting states, leading to a variety of novel phenomena. As a result, low-dimensional superconductors can exhibit phenomena that are weak or absent in bulk systems, including Berezinskii–Kosterlitz–Thouless behavior,[1,2] gate-tunable superconducting domes, anomalous metallic states, and possible topological superconductivity.[3–5] Recent progress in graphene-based moiré systems, transition-metal dichalcogenides (TMDs), FeSe-based interface superconductors, and hybrid topological platforms has therefore established low-dimensional superconductivity as a frontier that connects fundamental condensed-matter physics with emerging quantum-device concepts.[5,6]

In this chapter, the term low-dimensional superconducting materials is used in a broad sense to include purely two-dimensional (2D) materials, quasi-1D systems, superconducting interfaces, and moiré superlattice platforms. The scope is defined not merely by geometric thickness but, more importantly, by the effective dimensionality that governs superconducting behavior. Whereas in normal electronic transport, the relevant comparison is between the sample size and the inelastic scattering length, in superconductors, the key length scale is the superconducting coherence length ξ. A superconducting film or wire is therefore quasi-2D or quasi-1D when its thickness or width is smaller than or comparable to the superconducting coherence length ξ. Under this condition, reduced dimensionality has a profound influence on superconducting pairing, phase coherence, fluctuation effects, and external controllability.[4]

One of the most important conceptual differences between bulk and low-dimensional superconductors is that the superconducting transition is often governed not only by pair formation but also by the establishment of long-range or

DOI: 10.1201/9781003655305-7

quasi-long-range phase coherence. In reduced dimensions, phase stiffness is generally lower and thermal or quantum fluctuations are more effective, so the ordering temperature can be controlled as much by phase dynamics as by the microscopic pairing scale.[1,2] Reduced dimensionality also weakens electrostatic screening and enhances the impact of external perturbations. Electric gating, substrate charge transfer, strain, and interfacial hybridization can therefore tune carrier density, band dispersion, and even pairing tendencies much more efficiently than in conventional bulk superconductors. This high tunability is one of the key advantages of low-dimensional superconducting platforms.

At the same time, the effects of broken inversion symmetry, strong spin–orbit coupling, valley physics, and nontrivial band topology are often particularly pronounced in these systems. These ingredients can generate anisotropic critical fields, spin-momentum locking, unconventional pairing channels, and topological superconducting phases, thereby making reduced-dimensional systems a natural platform in which superconductivity, symmetry, and topology strongly intertwine.[5,7] Representative examples include unconventional superconducting states in moiré graphene and moiré TMD systems, chiral superconductivity in rhombohedral graphene, Ising superconductivity in monolayer TMDs, interfacial pairing modified by inversion breaking and Rashba-type spin–orbit coupling, and the broader search for topological superconductivity supporting protected boundary modes.[6,8] These aspects will recur throughout the remainder of the chapter. Against this conceptual background, several material families have emerged as representative platforms for low-dimensional superconductivity, each highlighting a different interplay among reduced dimensionality, symmetry, interactions, and tunability.

5.2 MAIN TYPES OF LOW-DIMENSIONAL SUPERCONDUCTING MATERIALS

Low-dimensional superconductivity has been realized in a wide range of material platforms, including atomically thin crystals, interfacial structures, moiré superlattices, and quasi-1D systems. Although these materials differ substantially in microscopic composition and pairing environment, they share the common feature that reduced dimensionality qualitatively modifies superconducting behavior through enhanced fluctuations, altered electronic structure, interfacial effects, and increased sensitivity to symmetry and external tuning. The study of low-dimensional superconductors is therefore not limited to identifying new superconducting materials but also aims to understand the distinct physical mechanisms that become prominent in reduced dimensions. In this section, we survey the principal material classes that have shaped this field and summarize their characteristic superconducting properties.

5.2.1 Superconductivity in Graphene and Its Derivatives

Pristine monolayer graphene is not, under ordinary conditions, a conventional superconductor. Nevertheless, graphene has become one of the most influential low-dimensional superconducting platforms because its band structure is exceptionally tunable and because superconductivity can be induced or engineered through multiple routes, including electrostatic doping, chemical intercalation, proximity coupling, multilayer stacking, and moiré band formation.[6]

A major conceptual shift occurred with the emergence of twisted bilayer and other moiré graphene systems. Near special twist angles, the moiré superlattice generates narrow electronic bands with strongly reduced kinetic energy, thereby enhancing the relative importance of Coulomb interactions. This creates an environment in which correlated insulating states, symmetry-broken metallic phases, and superconducting domes can appear in close proximity. As a result, moiré graphene has become a paradigmatic example of how band engineering and reduced dimensionality can produce superconductivity in a highly tunable correlated setting.[6,9] Another important development is the emergence of superconductivity in rhombohedral graphene systems. Owing to their extremely flat low-energy bands, these materials provide a natural setting for enhanced interaction effects and unconventional superconducting instabilities. Recent experiments have reported signatures consistent with chiral superconductivity in rhombohedral graphene, suggesting that graphene-based superconductivity can arise not only in moiré superlattices, but also in multilayer systems whose flat bands originate directly from stacking geometry itself.[10] This extends the scope of graphene-derived superconductivity and further strengthens its connection to topology, symmetry breaking, and correlation physics.

Graphene-based superconductors are especially valuable because they offer a particularly clear platform for exploring the interplay among superconductivity, topology, valley structure, and correlation effects. At the same time, the microscopic pairing mechanism remains the subject of active debate. Competing scenarios based on phonon-mediated attraction, electronically driven pairing, or a combination of several channels continue to be discussed, and this unresolved question is part of what makes graphene-derived superconductivity such an active frontier.[6]

5.2.2 Superconductivity in TMDs

TMDs provide one of the clearest demonstrations that reduced dimensionality and strong spin–orbit coupling can qualitatively reshape superconductivity. Metallic TMDs such as $NbSe_2$, NbS_2, TaS_2, and related compounds retain superconductivity down to the monolayer or few-layer limit, while simultaneously exhibiting strong anisotropy, proximity to charge-density-wave order, and pronounced sensitivity to symmetry breaking.[11] In these systems, the competition or coexistence between superconductivity and charge ordering is an important aspect of the low-energy physics. Reducing the thickness can alter the balance between these tendencies by changing screening, phonon spectra, and interlayer coupling, so that the thin-layer limit is not merely a weaker version of the bulk but a distinct electronic regime with its own hierarchy of instabilities.[12]

A particularly striking phenomenon is Ising superconductivity.[13,14] In monolayer TMDs lacking in-plane inversion symmetry, strong spin–orbit coupling can lock electron spins to out-of-plane directions near different valleys. This locking suppresses ordinary in-plane Zeeman depairing and can produce in-plane upper critical fields far above the conventional Pauli limit. TMDs therefore provide a particularly clear example of how symmetry and spin–orbit coupling can protect superconductivity in reduced dimensions.

More recently, superconductivity has also become an important frontier in moiré TMD systems, especially twisted TMD homobilayers such as twisted $MoTe_2$ and twisted WSe_2. In these materials, moiré band formation can generate narrow topological bands in which Coulomb interactions dominate over kinetic energy, while spin-valley locking and valley-contrasting Berry curvature further enrich the low-energy physics. Theory and recent experiments indicate that superconductivity in these systems can arise alongside a broad set of correlated and topological phases, including integer and fractional Chern insulating states, anomalous Hall states, and other interaction-driven orders.[15,16] Twisted TMDs therefore extend the superconducting landscape of this material family from the symmetry-protected physics of monolayer Ising superconductors to a regime where superconductivity may be closely intertwined with flat-band topology and nearby fractionalized states.

5.2.3 Topological Superconducting Materials

In addition to topological superconducting tendencies that may arise in specific material families such as iron chalcogenides, low-dimensional topological superconductivity is also extensively pursued in engineered platforms. In these systems, the central goal is to realize a superconducting phase with nontrivial topology that can support protected boundary excitations, most notably Majorana zero modes.[5,7] Reduced dimensionality is especially advantageous here because it facilitates interface design, electrostatic tunability, and the spatial isolation of edge, end, or vortex-bound states.[17,18] For this reason, low-dimensional heterostructures have become one of the main routes toward topological superconductivity.

One major class of such systems is semiconductor-superconductor hybrid structures. On these platforms, conventional s-wave superconductivity is combined with strong spin–orbit coupling and, typically, broken time-reversal symmetry induced by a magnetic field or a magnetic layer. Under suitable conditions, this combination can generate an effective topological superconducting phase in a 1D or quasi-1D setting.[18] These hybrids are especially important because they offer a high degree of external control and a relatively direct connection between microscopic design principles and device realization.

A second major route is provided by topological-material/superconductor interfaces, where superconductivity is induced into a system with nontrivial band topology through the proximity effect. This strategy is attractive because it combines the robust surface or edge states of topological materials with the pairing correlations of an adjacent superconductor, thereby creating a promising setting for topological superconductivity in reduced dimensions.[19,20] More generally, interface engineering has emerged as a unifying principle in this field; rather than relying solely on intrinsic materials, one can design topological superconducting behavior by combining complementary ingredients across a hybrid structure.[17,19]

A third important direction is magnet-superconductor hybrid systems, including magnetic chains, adatom arrays, and related atomically engineered structures on superconducting substrates. These platforms provide a conceptually distinct route to topological superconductivity, because magnetism itself helps reshape the low-energy quasiparticle spectrum into a topological phase.[21,22] Such systems are particularly appealing for local spectroscopic probes and for studying boundary modes at the atomic scale.

At the same time, topological superconductivity in real materials remains an active and sometimes controversial topic. Many experimental signatures are suggestive rather than definitive, and it is often challenging to distinguish genuine topological boundary states from trivial low-energy bound states. For this reason, the systems discussed here are best regarded as candidate platforms whose importance lies both in their fundamental promise and in the new materials-by-design strategies they have inspired.

5.2.4 Iron-Based Superconducting Materials

Iron-based superconductors form another important class of low-dimensional superconducting materials, especially in the form of ultrathin films, interfaces, and layered heterostructures. In this family, reduced dimensionality does not simply suppress bulk superconductivity, but can instead qualitatively reconstruct the superconducting state through interfacial charge transfer, modified electronic structure, enhanced anisotropy, and coupling to the substrate.[3] As a result, low-dimensional iron-based superconductors have become a central platform for studying how confinement and interface engineering can reshape pairing and enhance superconducting properties.

The most influential example is monolayer FeSe grown on $SrTiO_3$, where superconductivity is dramatically enhanced compared with bulk FeSe. This system has attracted enormous interest because it demonstrates that the interface itself can play an essential role in promoting superconductivity, through a combination of charge transfer, modified screening, substrate coupling, and interfacial phonon effects.[3] More broadly, FeSe-based thin films illustrate that the monolayer limit may constitute a distinct superconducting regime rather than a simple continuation of the bulk material.[3,23]

Another important aspect of low-dimensional iron-based superconductors is that some members of this family are also closely connected to topological superconductivity. In particular, iron chalcogenides such as Fe(Te, Se), as well as monolayer $FeTe_{1-x}Se_x$, have been widely discussed as candidate platforms in which superconductivity coexists with topological electronic states, thereby linking the iron-based family to the broader search for Majorana modes and topological superconducting phases.[17,24] In this sense, iron-based superconductors are important not only for their tunable and interface-enhanced pairing behavior, but also for providing a bridge between unconventional superconductivity and topology.

5.2.5 Other Low-Dimensional Superconducting Systems

In addition to the major material families discussed above, low-dimensional superconductivity also appears in a broader range of systems, including quasi-1D superconductors, oxide interfaces, and artificially engineered heterostructures. Although these platforms differ substantially in microscopic origin, they further demonstrate that superconductivity in reduced dimensions is often governed by a delicate interplay among confinement, fluctuations, disorder, interfacial coupling, and electronic reconstruction.

Quasi-1D superconducting systems provide a particularly instructive example. In such materials, the reduced transverse dimension enhances fluctuation effects and weakens long-range phase coherence, so that superconductivity is often more fragile than in 2D or bulk systems. Superconducting nanowires such as MoGe and ultrathin NbN structures, as well as related atomic-scale or proximitized 1D platforms, have served as important systems for studying phase slips, disorder effects, finite-size behavior, and the crossover between mesoscopic and macroscopic superconductivity.[25,26] These systems are therefore valuable for understanding how superconducting order evolves when spatial confinement becomes extreme.

Another major category is interfacial and oxide superconductivity. In these systems, superconductivity may emerge at a boundary between two materials even when neither constituent alone is an ordinary low-dimensional superconductor. Representative examples include the $LaAlO_3/SrTiO_3$ interface and superconducting 2D electron systems formed at $KTaO_3$ surfaces or interfaces.[27,28] Such platforms highlight the roles of charge transfer, orbital reconstruction, broken symmetry, and electronic confinement, and show that low-dimensional superconductivity can arise through interfacial electronic reconstruction rather than only from an intrinsically layered crystal structure.

Artificially assembled van der Waals heterostructures and multilayer superconducting devices provide yet another route to low-dimensional superconductivity. By stacking different atomically thin materials, one can design hybrid systems in which superconductivity coexists or competes with magnetism, strong spin–orbit coupling, topological band features, or correlated insulating behavior. These engineered structures are particularly appealing because they offer a high level of external control and design flexibility, making them natural platforms for studying proximity effects, tunable pairing environments, and novel superconducting device concepts.

Taken together, these additional systems emphasize that low-dimensional superconductivity should not be viewed as a narrow materials category, but rather as a broader organizing principle for understanding how superconducting order behaves under strong spatial confinement, interfacial reconstruction, and structural design. They also reinforce a central theme of this chapter: reduced dimensionality not only modifies known superconductors but also creates opportunities for qualitatively new superconducting phenomena and functionalities.

5.3 REGULATION OF LOW-DIMENSIONAL SUPERCONDUCTING MATERIALS

A defining feature of low-dimensional superconductors is not only their reduced dimensionality, but also the remarkable degree to which their superconducting properties can be tuned by external perturbations. Because the relevant electronic states are confined to a small thickness or an interfacial region, electric fields, magnetic fields, strain, substrate coupling, and chemical modification can strongly influence carrier density, band dispersion, spin texture, screening, and phase stiffness. As a result, superconductivity in low-dimensional systems is often best understood not as a fixed material property, but as a continuously tunable quantum state.

5.3.1 Electric Field Modulation

Electric field modulation is one of the important advantages of low-dimensional superconducting materials. In bulk superconductors, electrostatic control is often weak because metallic screening strongly suppresses field penetration. By contrast, atomically thin crystals, interface systems, and van der Waals heterostructures can respond dramatically to gating, allowing superconductivity to be induced, enhanced, suppressed, or continuously tuned over a wide parameter range. This makes the electric field not merely a probe, but a genuine thermodynamic control parameter of superconducting order. The most direct effect of gating is to vary the carrier density and therefore shift the Fermi level and modify the density of states and the occupation of relevant bands. In graphene-based systems and moiré materials, this can move the system between correlated insulating states, metallic phases, and superconducting domes. In TMDs and oxide interfaces, electrostatic or ionic gating can tune both transition temperature and critical fields, while in some semiconducting or weakly metallic materials, it can even trigger superconductivity from a non-superconducting parent state. These examples show that low-dimensional superconductivity is often embedded in a broader phase diagram that becomes accessible through field-effect control. Electric field modulation can also influence superconductivity more indirectly through screening, disorder, inversion-symmetry breaking, and interfacial electrostatics. Ionic-liquid or electrochemical gating, although less purely reversible than conventional solid-state gating, can reach much larger carrier concentrations and sometimes modify lattice or chemical environments in addition to the electronic density.[28,29]

Electric field modulation is especially effective in low-dimensional superconductors because the relevant electronic states are confined to an atomically thin layer, a narrow channel, or an interfacial region, so the applied field can directly reshape the low-energy electronic structure rather than being almost completely screened as in bulk metals. As a result, gating can do much more than simply change the carrier density. It can modify band dispersion, shift van Hove singularities, redistribute charge between layers, alter the Fermi-surface geometry, and change the balance between kinetic energy and Coulomb interaction. In moiré systems, this is particularly important because gate control can shift the system across different integer and fractional fillings of narrow bands, while a displacement field can additionally tune layer polarization and the topology of the active bands. Twisted homobilayer TMDs provide a clear example: their phase diagrams are highly sensitive to carrier density, displacement field, and twist angle, and include superconductivity alongside integer and fractional topological phases.[15] Rhombohedral trilayer graphene and magic-angle twisted trilayer graphene provide further examples in which the displacement field acts as an essential tuning parameter for correlated and superconducting phases.[30,31]

It is also useful to distinguish between electrostatic gating and ionic gating. Conventional electrostatic gates mainly tune the chemical potential and displacement field in a reversible and comparatively clean way, making them particularly well-suited for graphene-based moiré materials and dual-gated van der Waals devices. Ionic gating, by contrast, can generate much larger interfacial electric fields and much higher induced carrier densities through electric double-layer formation, and has therefore become a powerful route to driving or strongly enhancing superconductivity in thin films and oxide surfaces.[29] A representative example is the induction of 2D superconductivity at $KTaO_3$ surfaces by ionic-liquid gating, which illustrates how electric field control can create superconductivity in a system that is not superconducting in its ungated surface state.[28]

Representative examples also show that electric fields in low-dimensional systems act not merely on the carrier density, but also on the interaction strength and Fermi-surface structure that underlie superconducting pairing. In rhombohedral trilayer graphene, superconductivity was observed in a phase diagram controlled jointly by carrier density and electric displacement field, with two distinct superconducting regions appearing in gate-tuned parameter space.[30] In magic-angle twisted bilayer graphene, screening by a nearby Bernal bilayer can continuously tune the effective Coulomb interaction, and this control has been used to probe the origin of the superconducting state.[32] In magic-angle twisted trilayer graphene, superconductivity is likewise strongly displacement-field tunable, and recent work has reported double-dome superconductivity and competing magnetic order within a finite displacement-field window.[31] Beyond tuning transition temperatures and superconducting domes, electric fields can also control more subtle aspects of the superconducting state, including phase stiffness, layer polarization, and inversion symmetry. This broader control is important because superconductivity in low-dimensional systems is often intertwined with nearby topological and symmetry-broken phases. For example, gate-defined weak links in magic-angle twisted bilayer graphene have been used to realize symmetry-broken Josephson junctions and superconducting-diode behavior.[33]

More generally, electric field control can tune the diode efficiency in multiterminal Josephson devices,[34] while inversion-symmetry breaking in low-dimensional superconductors, such as stepped $NbSe_2$ nanoflakes, can induce superconducting-diode behavior.[35] In this sense, the electric field acts not only on the charge sector, but on the entire correlated and symmetry-sensitive environment in which superconductivity develops.

5.3.2 Magnetic Field Modulation

Magnetic-field modulation is another central route for controlling low-dimensional superconductivity. In contrast to electric gating, which primarily acts through carrier density and electrostatics, the magnetic field couples directly to both the orbital motion and the spin degree of freedom of Cooper pairs. Because orbital motion is strongly constrained in reduced dimensions, the response to magnetic field is often highly anisotropic: out-of-plane fields efficiently generate vortices and suppress phase coherence, whereas in-plane fields can probe Zeeman depairing more directly. As a result, the magnetic field provides a particularly powerful means of distinguishing the roles of orbital pair breaking, spin–orbit coupling, vortex physics, and quantum fluctuations in low-dimensional superconducting materials.[36]

A striking example is provided by TMDs, where monolayer or few-layer samples can exhibit Ising superconductivity. In these systems, broken in-plane inversion symmetry together with strong spin–orbit coupling locks electron spins predominantly out of plane near opposite valleys, thereby strongly suppressing ordinary in-plane Zeeman depairing. As a consequence, the in-plane upper critical field can far exceed the conventional Pauli limit, making TMDs a textbook platform for magnetic-field-resilient superconductivity in reduced dimensions.[13,37] This behavior has been observed in several atomically thin TMD superconductors and has established magnetic-field anisotropy as one of the clearest signatures that superconductivity in the 2D limit can differ qualitatively from its bulk counterpart.

Magnetic field is also indispensable for controlling vortex physics in low-dimensional superconductors. In a perpendicular field, vortices enter the sample, and their density, mobility, and mutual interactions govern dissipation and phase coherence. In 2D systems, this often leads to rich field-tuned behavior, including vortex liquid regimes, anomalous metallic states, and magnetic-field-driven superconductor-insulator transitions (SITs).[36,38] Highly crystalline 2D superconductors are particularly revealing in this regard, because reduced disorder allows field-driven quantum phase transitions to be studied in a regime where vortex motion and quantum fluctuations play a dominant role.[36] Thus, the magnetic field is not only a depairing mechanism but also a direct means of tuning the topology and dynamics of the superconducting condensate by creating and manipulating vortices.

Beyond suppressing superconductivity, a magnetic field can also induce or enhance more subtle nonreciprocal effects. In low-dimensional systems with strong spin–orbit coupling or broken inversion symmetry, a magnetic field can cooperate with structural asymmetry to generate superconducting diode behavior, namely a directional asymmetry in the critical current. Few-layer $NbSe_2$ provides a representative example: field-tunable diode effects have been reported both in stepped nanoflakes, where inversion symmetry is broken by thickness asymmetry and in devices where spin–orbit coupling and symmetry breaking combine to generate a zero-field or weak-field nonreciprocal response.[35,39] More generally, the magnetic field has become a key experimental knob for diagnosing symmetry breaking and nonreciprocal transport in low-dimensional superconducting devices, linking field control not only to phase transitions and critical fields but also to emergent functionalities relevant to superconducting electronics.[40]

5.3.3 Strain Modulation

Strain modulation is especially effective in low-dimensional materials because thin crystals and heterostructures are mechanically flexible and highly sensitive to small structural distortions. Unlike a uniform field that acts primarily on charge or spin, strain directly reshapes bond lengths, hopping amplitudes, crystal fields, lattice symmetry, and phonon spectra. Since superconductivity depends sensitively on the underlying band structure and pairing environment, even modest strain can significantly change transition temperatures, gap structure, or the competition between superconductivity and neighboring ordered states. In graphene-based and moiré systems, strain can alter band flatness, local stacking configurations, and the relative stability of correlated phases. In TMDs and FeSe-based films, strain may affect electron-phonon coupling, Fermi-surface geometry, and interfacial interactions.

It is therefore useful to regard strain as more than a secondary perturbation. In many low-dimensional superconductors, it is a genuine design tool that can redistribute spectral weight, modify symmetry, and tune the balance between itinerancy and correlation.[41,42]

In graphene-based moiré materials, strain is especially important because the relevant electronic structure is itself generated by geometric interference between layers. Even a weak heterostrain can substantially modify the bandwidth, valley structure, and separation of flat bands in twisted bilayer graphene, thereby changing the energetic balance among metallic, correlated insulating, and superconducting states.[41,43] More broadly, strain engineering in twisted graphene has become a powerful route for tuning moiré reconstruction, local stacking, and flat-band formation and can therefore reshape the superconducting phase diagram at a very basic level.[43] In this sense, strain acts not merely as a perturbation on an existing superconducting state, but as a parameter that helps define the correlated electronic landscape from which superconductivity emerges.

In TMDs, strain provides a complementary route by tuning both electronic dispersion and lattice dynamics. Because superconductivity in materials such as $NbSe_2$ develops in proximity to charge-density-wave order, strain can strongly influence the competition between these two instabilities. Small uniaxial strain has been shown to produce substantial changes in the electronic structure of 2H-$NbSe_2$ and to lower its symmetry,[44] while atomic-scale strain engineering can induce different charge-density-wave geometries in $NbSe_2$, demonstrating the remarkable sensitivity of this material to lattice distortion.[45] Substrate proximity can also modify charge-density-wave and superconducting behavior in single-layer $NbSe_2$, reinforcing the broader point that, in low-dimensional TMDs, superconductivity is closely tied to strain, environment, and lattice-coupled competing orders.[46] These examples show that strain can tune superconductivity not only through the density of states but also through phonons and neighboring ordered phases.

FeSe-based systems provide another important setting in which strain is deeply intertwined with superconductivity. In monolayer FeSe on $SrTiO_3$, the superconducting state is strongly affected by the interface, and early work showed that strain is closely linked to the suppression of spin-density-wave tendencies as thickness increases or epitaxial strain is relaxed.[42] More recent work on multilayer FeSe/$SrTiO_3$ films has further shown that tensile strain can suppress nematicity near the interface.[47] More generally, FeSe-based thin films illustrate how strain, interfacial charge transfer, and substrate coupling can work together to reshape the Fermi surface and pairing environment. Strain modulation is therefore particularly valuable in these systems because it provides a route to tune superconductivity indirectly through its coupling to magnetism, nematicity, and interfacial electronic reconstruction.

Taken together, these examples show that strain modulation in low-dimensional superconductors operates through multiple intertwined channels: it can alter band dispersion, shift van Hove singularities, modify phonon spectra, change interlayer coupling, lower lattice symmetry, and tune the balance between superconductivity and competing orders. Compared with electric and magnetic fields, strain is distinctive in that it acts directly on the microscopic structure from which the low-energy electronic problem arises. For this reason, strain engineering has become an increasingly important route not only for optimizing superconducting properties but also for designing new low-dimensional superconducting phases and devices.[41,43]

5.3.4 Other Regulation Methods

In addition to the electric field, magnetic field, and strain, a wider set of tuning approaches is important for understanding and controlling low-dimensional superconductivity. These approaches are diverse in experimental implementation, but they can be understood more clearly by asking what they modify physically. Some primarily change bandwidth, interlayer spacing, and crystal structure, as in the case of pressure and twist-angle control. Others mainly tune interface coupling, charge transfer, and symmetry, as in substrate engineering. Chemical intercalation acts more directly on carrier density, screening, and electron-phonon coupling, whereas defect engineering controls the degree of disorder, scattering, and inhomogeneity. Nonequilibrium optical excitation, finally, offers a route to transiently reshape the electronic or lattice environment on ultrafast timescales. Viewed in this way, these methods collectively reinforce a central lesson of low-dimensional superconductivity: the superconducting state is unusually sensitive to modifications of its microscopic environment and can therefore be controlled through multiple intertwined channels.

One important group of methods acts primarily by modifying bandwidth, lattice geometry, and interlayer coupling. Pressure changes bond lengths, orbital overlap, and interlayer spacing, thereby reshaping the electronic structure and the balance among superconductivity, magnetism, and nematicity. FeSe is a classic example, where pressure strongly reconstructs the superconducting phase diagram through its coupling to competing orders.[48] In moiré materials, an analogous but uniquely low-dimensional tuning knob is the twist angle itself: varying the relative rotational alignment between layers can transform dispersive bands into ultra-flat or topological bands, thereby reconstructing the correlated phase diagram from the outset rather than merely shifting an already existing superconducting state. Twist-angle control, therefore, differs qualitatively from ordinary tuning in bulk materials, because here the geometric arrangement of the lattice becomes a primary control variable for superconductivity.

A second group of methods mainly changes interface coupling, charge transfer, screening, and pairing environment. Substrate engineering is particularly powerful in low-dimensional systems because the active electronic states often lie directly at or near an interface. The best-known example is monolayer FeSe on $SrTiO_3$, where superconductivity is substantially enhanced relative to bulk FeSe and where interfacial coupling, charge transfer, and substrate phonons have all been implicated in shaping the superconducting state.[49] Chemical doping and intercalation operate through a related but more direct channel. By inserting atoms or molecules between layers or at interfaces, one can inject carriers, alter screening, and in some cases induce superconductivity altogether. Graphene-based systems provide a representative example, as superconductivity has been observed in calcium-intercalated bilayer graphene and related intercalated graphene structures.[50,51]

A third group of methods acts primarily on disorder, local symmetry breaking, and dynamical nonequilibrium effects. Defect engineering is important because reduced dimensionality makes superconductivity especially sensitive to vacancies, substitutional disorder, and local structural distortions. In monolayer or few-layer $NbSe_2$, defects and vacancy-related distortions can modify both local symmetry and interlayer coupling, thereby influencing superconducting behavior and its interplay with charge ordering.[52,53] Optical nonequilibrium control represents a different limit, where the goal is not static tuning but ultrafast transient manipulation. Recent work has shown reversible optical modulation of superconductivity in thin-film $NbSe_2$ via plasmonic near-field coupling,[54] while broader ultrafast studies have highlighted light as a means of transiently reshaping superconducting and competing orders on femtosecond timescales.[55]

Although these approaches are less established as equilibrium tuning methods, they point toward a future in which low-dimensional superconductivity can be steered not only statically but also dynamically. Taken together, these additional regulation methods show that low-dimensional superconductivity is best understood as a multidimensional control problem. Some tuning knobs primarily modify carrier density and screening, others reshape lattice geometry and bandwidth, others act through interfaces and symmetry, and still others control disorder or drive the system out of equilibrium. This perspective helps unify a wide range of experimental techniques within a single physical framework and underscores why low-dimensional superconductors are such fertile systems for both fundamental studies and materials-by-design strategies.

5.4 QUANTUM PHASE TRANSITIONS IN LOW-DIMENSIONAL SUPERCONDUCTING MATERIALS

Quantum phase transitions are a natural and important focus in low-dimensional superconductors because these systems combine strong fluctuations with experimentally accessible nonthermal tuning parameters such as carrier density, magnetic field, thickness, strain, and disorder. Unlike a thermal phase transition, which is driven by temperature, a quantum phase transition is driven by variation of a control parameter in the zero-temperature, or experimentally, the very low-temperature limit. Low-dimensional platforms are particularly favorable for observing such phenomena, since reduced phase stiffness, enhanced fluctuation effects, and greater sensitivity to inhomogeneity often render the superconducting state comparatively fragile and thus more readily destabilized under controlled conditions. In this section, we discuss three representative classes of quantum phase transitions, including the emergence of superconductivity from a metallic state, the destruction of superconductivity into an insulating state, and topological transitions between distinct superconducting phases.

5.4.1 Metal-Superconductor Phase Transition

The metal-superconductor transition in low-dimensional materials naturally falls into the category of quantum phase transitions when it is tuned by carrier density, thickness, disorder, pressure, or magnetic field. Because reduced dimensionality enhances fluctuations and weakens phase stiffness, the onset of superconductivity in these systems is often more delicate than in bulk materials and can be tuned continuously through experimentally accessible parameters.[4]

Experimentally, the superconducting onset can be driven in several ways. Gate tuning is especially powerful in interfacial and atomically thin systems, where increasing carrier density can stabilize superconductivity by moving the Fermi level into a more favorable regime and by modifying screening and disorder. A representative example is the $LaAlO_3/SrTiO_3$ interface, where dual gating has been used to control carrier density and disorder independently, revealing a 2D superconducting dome and a gate-tuned superconductor-metal transition.[56]

Thickness control provides another classic route: in ultrathin films, reducing thickness enhances confinement, disorder, and fluctuations, often suppressing superconductivity or shifting its onset in a systematic way. For example, thickness-driven superconducting evolution has been reported in ultrathin NbS_2 and in granular ultrathin NbN films, illustrating how geometric confinement alone can push the system toward the edge of superconductivity.[26,57] These examples show that the metallic-to-superconducting boundary in low-dimensional systems is often accessible through controlled modification of carrier density, confinement, and disorder.

A central conceptual point is that the transition is governed not only by the development of pairing correlations but also by the establishment of global phase coherence. In low-dimensional materials, these two scales can separate; local Cooper pairing may survive even when the superfluid stiffness is too small to support a fully coherent zero-resistance state. This distinction has been emphasized in oxide-interface superconductors, where dual-gate measurements indicate that precursor pairing and long-range phase coherence can evolve differently near the transition.[56] More broadly, reviews of quantum phase transitions in 2D superconductors stress that reduced phase stiffness and enhanced inhomogeneity can make the superconducting state fragile, so that the destruction or recovery of coherence becomes as important as pairing amplitude itself.[4] This is one reason why the metal-superconductor boundary in low dimensions often appears broadened or fluctuation-dominated rather than sharply mean-field-like.

An important feature of this transition is the possible appearance of an anomalous metallic regime near the superconducting boundary. In such a state, the resistance drops far below the normal-state value but saturates to a finite value instead of vanishing as temperature approaches zero. This behavior has now been observed in a wide range of low-dimensional superconducting systems, including ion-gated crystalline superconductors, oxide interfaces, and disordered films.[58,59] Although its microscopic origin remains actively debated, the anomalous metal is widely regarded as evidence that the boundary between metal and superconductor in low-dimensional systems can broaden into a distinct low-temperature regime rather than remain a sharp classical divide. In this sense, the metal-superconductor transition in reduced dimensions is often best viewed not as a simple binary switch, but as a fluctuation-rich quantum evolution in which pairing, coherence, dissipation, and inhomogeneity may all play essential roles. If the first type of quantum transition concerns how superconductivity emerges from a metallic background, a second and equally important question is how a superconducting state is destroyed altogether. In low-dimensional systems, this often leads not back to an ordinary metal, but toward an insulating phase shaped by localization, disorder, and quantum fluctuations.

5.4.2 Superconductor-Insulator Phase Transition

The SIT is one of the most important quantum phase transitions in low-dimensional superconductors. Unlike the metal-superconductor transition, which concerns the onset of a zero-resistance state from a metallic normal phase, the SIT describes the destruction of superconductivity in favor of an insulating ground state as a nonthermal control parameter is varied. In 2D systems, this transition is widely regarded as a prototypical quantum phase transition driven by quantum fluctuations in the zero-temperature limit.[4,36] Low-dimensional systems are especially favorable for observing SIT behavior because reduced dimensionality enhances phase fluctuations, disorder effects, and electronic inhomogeneity, all of which can destabilize superconductivity even when local pairing correlations remain significant.[4,60]

Experimentally, the SIT can be driven by several control parameters, most prominently magnetic field, disorder, and thickness. In highly disordered or ultrathin films, increasing disorder or reducing thickness can suppress superconductivity and drive the system into an insulating state.[26,60] A perpendicular magnetic field provides another widely used tuning knob by generating vortices, enhancing phase frustration, and eventually destroying long-range coherence; field-driven SITs have been reported in numerous thin-film systems and remain a central framework for studying quantum criticality in 2D superconductors.[36,38] In this sense, the SIT is not simply the destruction of superconductivity, but a competition between superconducting coherence, localization, and interaction effects, with the balance among them controlled by experimentally accessible parameters.

A central conceptual issue in the SIT is whether the transition is best viewed in a bosonic or fermionic picture. In a bosonic scenario, Cooper pairs survive across the transition but lose global phase coherence, so the insulating state retains localized pairs. In a fermionic scenario, by contrast, the pairing amplitude itself is destroyed, and the system crosses over to a more conventional electronic insulator. Experimental studies continue to find evidence relevant to both viewpoints, and the answer may depend strongly on disorder, homogeneity, and the specific material platform.[4,61] This distinction is important because it shows that the insulating phase reached from a low-dimensional superconductor need not be a trivial normal insulator, but may still carry remnants of superconducting correlations.

Another important development is the recognition that the SIT is often accompanied by or interrupted by an anomalous metallic regime, especially in cleaner or weakly disordered 2D systems. In such cases, the resistance saturates to a finite low-temperature value over a range of tuning parameters, rather than diverging as in a conventional insulator or vanishing as in a true superconductor.[36,58,60] This has led to the view that the superconductor-insulator boundary in low-dimensional materials may broaden into a richer quantum landscape involving anomalous metals, quantum Griffiths behavior, and broadened critical regimes rather than a single sharp transition. The SIT in reduced dimensions should therefore be understood not only as a canonical example of quantum criticality but also as a testing ground for the interplay among superconducting pairing, coherence, localization, and dissipation.[58] The two transitions discussed above primarily concern whether superconductivity appears or disappears. A further possibility, however, is that superconductivity remains intact while its internal topological character changes. This leads to a third class of quantum phase transitions, namely, transitions between topologically distinct superconducting phases.

5.4.3 Topological Phase Transition

A topological phase transition differs from ordinary symmetry-breaking transitions in that the relevant distinction is not simply whether an order parameter is present, but whether the quantum state belongs to a topologically trivial or nontrivial phase. In superconducting systems, this usually refers to transition between a conventional superconducting state and a topological superconducting state that can host protected boundary excitations such as Majorana zero modes.[5,7] In low-dimensional platforms, such transitions are especially important because reduced dimensionality makes it easier to engineer the ingredients needed for topological superconductivity, including strong spin–orbit coupling, symmetry breaking, interface-induced pairing, and boundary confinement.[5,62] Topological phase transitions are typically accompanied by a closing and reopening of the bulk quasiparticle gap as an external control parameter is varied, thereby changing the corresponding topological invariant.[62,63]

One of the most influential examples is the proximitized semiconductor nanowire, where a magnetic field, together with spin–orbit coupling and induced s-wave pairing, can drive a transition from a trivial superconducting phase to a 1D topological superconducting phase.[18,63] In this setting, the relevant tuning parameters are the Zeeman field and the chemical potential, both of which can be controlled experimentally through magnetic field and gating. The basic physical picture is that once the Zeeman splitting exceeds a critical scale set by the induced gap and band filling, the bulk gap closes and the wire enters a topological regime with Majorana end states.[18,63] This has made nanowire platforms a paradigmatic setting for discussing topological phase transitions in a particularly transparent way.

A second important route is provided by topological-insulator/superconductor interfaces and related hybrid structures. In these systems, superconductivity is induced on topological surface or edge states through the proximity effect, and the topological character of the normal-state band structure plays an essential role in shaping the superconducting phase.[19] More generally, interface engineering has become a central strategy for realizing topological superconducting transitions in low-dimensional materials because changing the interfacial environment can tune pairing strength, spin texture, and effective symmetry breaking. Recent work on interface-induced superconductivity in magnetic topological materials further illustrates that combining superconductivity with topological and magnetic degrees of freedom can open new routes toward topological superconducting phases in reduced dimensions.[20]

More broadly, topological phase transitions in low-dimensional superconductors can be driven not only by magnetic field and chemical potential but also by interface coupling, symmetry breaking, and band-structure reconstruction. This is one reason why the search for topological superconductivity overlaps so strongly with the broader program of materials and heterostructure design. At the same time, identifying such transitions experimentally remains challenging, since candidate signatures such as zero-bias states may also arise from trivial low-energy bound states. For this reason, topological phase transitions in superconductors are best viewed not only as a route toward Majorana physics but also as a broader example of how low-dimensional superconductivity can be reshaped by topology, symmetry, and quantum engineering.[19,20,62]

5.5 APPLICATIONS OF LOW-DIMENSIONAL SUPERCONDUCTING MATERIALS

The application value of low-dimensional superconductors lies not only in the existence of superconductivity itself but also in the unusual tunability, interface compatibility, and design flexibility that reduced dimensionality makes possible. Because superconducting order in these systems can often be controlled by gating, magnetic field, strain, stacking geometry, and interface engineering, low-dimensional superconductors are attractive both as platforms for fundamental quantum phenomena and as ingredients for new device concepts. Their importance, therefore, does not simply come from realizing conventional superconducting functionality in thinner form, but from enabling device behavior that is difficult to achieve in bulk materials. In practice, the most important application directions currently lie in quantum information and in tunable superconducting electronics and sensing.[12]

5.5.1 Quantum Information and Topological Devices

One of the most intensively discussed application directions is quantum information. Low-dimensional superconducting platforms are particularly attractive in this context because they provide practical routes toward topological superconductivity and Majorana boundary modes, which have long been proposed as a basis for fault-tolerant or partially protected qubit architectures.[7] Semiconductor-superconductor nanowires, topological-insulator/superconductor interfaces, and magnet-superconductor hybrid systems are all examples in which reduced dimensionality helps isolate and manipulate boundary excitations that would be difficult to realize in bulk materials.[7] Even though the unambiguous identification of topological superconductivity remains challenging, the search itself has already strongly influenced the design of superconducting quantum materials and devices.

Low-dimensional superconductors are also relevant more broadly to superconducting quantum circuits. Their gate tunability, reduced size, and compatibility with hybrid heterostructures make them promising for flux-tunable circuit elements, superconducting weak links, and other miniaturized quantum components. Recent work has demonstrated graphene-based superconducting quantum circuits integrated into three-dimensional cavity architectures, showing that atomically thin materials can function as tunable superconducting elements in device geometries directly relevant to quantum hardware.[64] In this respect, the importance of low-dimensional superconductors is not limited to topological proposals alone. They also broaden the materials basis of superconducting quantum hardware by enabling the direct combination of superconductivity with low-dimensional semiconductors and topological materials.

5.5.2 Superconducting Electronics and Sensing

A second major application direction is superconducting electronics and sensing. Low-dimensional superconductors are well suited for this purpose because their critical currents, phase stiffness, and Josephson properties can often be tuned efficiently by external controls, making them promising materials for superconducting switches, weak links, reconfigurable Josephson elements, and low-power logic components.[12] In particular, gate-tunable Josephson junctions and nonreciprocal superconducting devices illustrate how reduced dimensionality enables transport functionality that is difficult to realize in conventional bulk-based superconducting circuits.[33,34]

Among these emerging applications, the superconducting diode effect deserves special emphasis. A superconducting diode allows supercurrent to flow more easily in one direction than in the other, thereby introducing nonreciprocal transport into a dissipationless circuit element.[40] This behavior is especially attractive for superconducting electronics because it may enable directional current control, rectification, and new circuit architectures without relying on conventional resistive elements. Low-dimensional materials are particularly promising for this purpose, since inversion-symmetry breaking, strong spin–orbit coupling, gate tunability, and interface engineering can all be used to generate and manipulate diode behavior. Experiments in multiterminal Josephson devices and magic-angle graphene junctions have already shown that diode efficiency and nonreciprocal transport can be tuned electrically, pointing toward possible applications in superconducting rectifiers, programmable circuit elements, and low-dissipation signal-processing devices.[33,34]

Low-dimensional superconductors are also promising for sensitive detectors and hybrid sensing platforms. Their small volume, large anisotropy, and strong response to external perturbations make them attractive for devices based on critical-current modulation, vortex dynamics, or phase-coherent interference. More broadly, 2D crystalline and interface superconductors are increasingly viewed as a platform in which sensing, switching, and correlated-state control can be combined within a single material system. This does not necessarily mean that low-dimensional superconductors will replace established superconducting technologies in the near term. Rather, their importance lies in enabling more compact, more tunable, and more multifunctional superconducting device architectures.

Taken together, these developments suggest that the applications of low-dimensional superconductors are currently most compelling in two closely connected directions, i.e., quantum information and tunable superconducting electronics. At the same time, important challenges remain, including materials uniformity, reproducibility, scalable fabrication, and the reliable identification of genuinely topological or interaction-driven device states.[5] The long-term impact of low-dimensional superconductors will therefore likely come not from a single application alone, but from the broader convergence of superconductivity, quantum materials design, and device-level controllability that reduced dimensionality uniquely enables.[12]

5.6 SUMMARY AND OUTLOOK

Low-dimensional superconducting materials have emerged as one of the most vibrant areas of modern quantum materials research. As discussed in this chapter, reduced dimensionality does not merely confine superconductivity to a smaller spatial scale but can fundamentally alter its behavior through enhanced fluctuations, weaker screening, stronger interfacial effects, and greater sensitivity to symmetry, topology, and external tuning. This has led to a broad landscape of superconducting platforms, including graphene-based systems, TMDs, iron-based thin films, topological superconducting heterostructures, and other interfacial or quasi-1D systems, each of which highlights a different aspect of how superconductivity evolves in reduced dimensions.

A central feature shared by these materials is their tunability. Electric field, magnetic field, strain, pressure, twist angle, substrate engineering, chemical modification, and defect control can all reshape superconducting behavior far more efficiently than in conventional bulk superconductors. This unusual controllability has made low-dimensional superconductors a particularly powerful setting for studying quantum phase transitions, competing ordered states, and the interplay among pairing, phase coherence, and topology. At the same time, it has opened the door to new application directions in quantum information, tunable superconducting electronics, and nonreciprocal superconducting devices such as superconducting diodes.

Looking ahead, several major challenges and opportunities remain. On the fundamental side, the microscopic pairing mechanisms in many low-dimensional superconductors, especially in moiré and other strongly correlated systems, remain poorly understood. The identification of unambiguous topological superconductivity also remains a central open problem. On the materials side, improved sample uniformity, reproducibility, and scalable fabrication will be essential if these systems are to move beyond proof-of-principle experiments toward robust device applications. Nevertheless, the rapid progress of recent years strongly suggests that low-dimensional superconductors will continue to play a central role in connecting quantum materials discovery with quantum-device design. Their long-term importance will likely lie not in any single material family alone, but in the broader realization that superconductivity in reduced dimensions is a highly controllable quantum state that can be engineered, tuned, and integrated in ways not possible in conventional bulk systems.

REFERENCES

[1] V. L. Berezinskii. Destruction of long-range order in one-dimensional and two-dimensional systems possessing a continuous symmetry group. II. Quantum systems. *Soviet Physics JETP*, 34, 610–616 (1972).

[2] J. M. Kosterlitz and D. J. Thouless. Ordering, metastability and phase transitions in two-dimensional systems. *Journal of Physics C: Solid State Physics*, 6, 1181 (1973).

[3] Joshua Maggiora, Xiaolin Wang, and Rongkun Zheng. Superconductivity and interfaces. *Physics Reports*, 1076, 1–49 (2024).

[4] Ziqiao Wang, Yi Liu, Chengcheng Ji, and Jian Wang. Quantum phase transitions in two-dimensional superconductors: A review on recent experimental progress. *Reports on Progress in Physics*, 87, 014502 (2023).

[5] Masatoshi Sato and Yoichi Ando. Topological superconductors: A review. *Reports on Progress in Physics*, 80, 076501 (2017).

[6] Leon Balents, Cory R. Dean, Dmitri K. Efetov, and Andrea F. Young. Superconductivity and strong correlations in moiré flat bands. *Nature Physics*, 16, 725–733 (2020).
[7] Jason Alicea. New directions in the pursuit of Majorana fermions in solid state systems. *Reports on Progress in Physics*, 75, 076501 (2012).
[8] Darshana Wickramaratne and I. I. Mazin. Ising superconductivity: A first-principles perspective. *Applied Physics Letters*, 122, 240503, (2023).
[9] Yuan Cao et al. Unconventional superconductivity in magic-angle graphene superlattices. *Nature*, 556, 43–50 (2018).
[10] Tonghang Han et al. Signatures of chiral superconductivity in rhombohedral graphene. *Nature*, 643, 654–661 (2025).
[11] Shahar Simon et al. The transition-metal-dichalcogenide family as a superconductor tuned by charge density wave strength. *Nature Communications*, 15, 10439 (2024).
[12] Haoran Ji, Yi Liu, Chengcheng Ji, and Jian Wang. Two-dimensional and interface superconductivity in crystalline systems. *Accounts of Materials Research*, 5, 1146–1157 (2024).
[13] Chong Wang, Yong Xu, and Wenhui Duan. Ising superconductivity and its hidden variants. *Accounts of Materials Research*, 2, 526–533 (2021).
[14] Haijiao Ji, Haiwen Liu, Hua Jiang, and X. C. Xie. Disorder effects on quantum transport and quantum phase transition in low-dimensional superconducting and topological systems. *Advances in Physics: X*, 6, 1884133 (2021).
[15] Bohao Li, Wen-Xuan Qiu, Fengcheng Wu, and A H MacDonald. Quantum phases in twisted homobilayer transition metal dichalcogenides. *National Science Review*, 13, nwaf 570 (2025).
[16] Jihan Zhu, Yang-Zhi Chou, Ming Xie, and Sankar Das Sarma. Superconductivity in twisted transition metal dichalcogenide homobilayers. *Physical Review B*, 111, L060501 (2025).
[17] Manasi Mandal et al. Topological superconductors from a materials perspective. *Chemistry of Materials*, 35, 6184–6200 (2023).
[18] Roman M. Lutchyn, Jay D. Sau, and S. Das Sarma. Majorana fermions and a topological phase transition in semiconductor-superconductor heterostructures. *Physical Review Letters*, 105, 077001 (2010).
[19] Liang Fu and C. L. Kane. Superconducting proximity effect and Majorana fermions at the surface of a topological insulator. *Physical Review Letters*, 100, 096407 (2008).
[20] Hemian Yi et al. Interface-induced superconductivity in magnetic topological insulators. *Science*, 383, 634–639 (2024).
[21] Stephan Rachel and Roland Wiesendanger. Majorana quasiparticles in atomic spin chains on superconductors. *Physics Reports*, 1099, 1–28 (2025).
[22] Stevan Nadj-Perge et al. Observation of Majorana fermions in ferromagnetic atomic chains on a superconductor. *Science*, 346, 602–607 (2014).
[23] Xiaotong Jiao et al. Significantly enhanced superconductivity in monolayer FeSe films on $SrTiO_3$(001) via metallic δ-doping. *National Science Review*, 11, nwad213 (2023).
[24] Xianxin Wu, Xin Liu, Ronny Thomale, and Chao-Xing Liu. High-T_c superconductor Fe(Se,Te) monolayer: An intrinsic, scalable and electrically tunable majorana platform. *National Science Review*, 9, nwab087 (2021).
[25] A. Bezryadin, C. N. Lau, and M. Tinkham. Quantum suppression of superconductivity in ultrathin nanowires. *Nature*, 404, 971–974 (2000).
[26] T.-Y. Jing, Z.-Y. Han, Z.-H. He, M.-X. Shao, P. Li, and Z.-Q. Li. Quantum phase transition in NbN superconducting thin films. *Physical Review B*, 107, 184515 (2023).
[27] N. Reyren et al. Superconducting interfaces between insulating oxides. *Science*, 317, 1196–1199 (2007).

[28] Tianshuang Ren et al. Two-dimensional superconductivity at the surfaces of $KTaO_3$ gated with ionic liquid. *Science Advances*, 8, eabn4273 (2022).
[29] Erik Piatti. Ionic gating in metallic superconductors: A brief review. *Nano Express*, 2, 024003 (2021).
[30] Haoxing Zhou, Tian Xie, Takashi Taniguchi, Kenji Watanabe, and Andrea F. Young. Superconductivity in rhombohedral trilayer graphene. *Nature*, 598, 434–438 (2021).
[31] Zekang Zhou et al. Gate-tunable double-dome superconductivity in twisted trilayer graphene. *Nature Physics*, 21, 1773–1779 (2025).
[32] Xiaoxue Liu, Zhi Wang, K. Watanabe, T. Taniguchi, Oskar Vafek, and J. I. A. Li. Tuning electron correlation in magic-angle twisted bilayer graphene using Coulomb screening. *Science*, 371, 1261–1265 (2021).
[33] J. Díez-Mérida et al. Symmetry-broken Josephson junctions and superconducting diodes in magic-angle twisted bilayer graphene. *Nature Communications*, 14, 2396 (2023).
[34] Mohit Gupta et al. Gate-tunable superconducting diode effect in a three-terminal Josephson device. *Nature Communications*, 14, 3078 (2023).
[35] Changlong Wang et al. Superconducting-diode effect induced by inversion-symmetry breaking in a stepped $NbSe_2$ nanoflake. *Physical Review Applied*, 22, 064017 (2024).
[36] Yu Saito, Tsutomu Nojima, and Yoshihiro Iwasa. Quantum phase transitions in highly crystalline two-dimensional superconductors. *Nature Communications*, 9, 778 (2018).
[37] Ying Xing et al. Ising superconductivity and quantum phase transition in macro-size monolayer $NbSe_2$. *Nano Letters*, 17, 6802–6807 (2017).
[38] Xiaofu Zhang et al. Size dependent nature of the magnetic-field driven superconductor-to-insulator quantum-phase transitions. *Communications Physics*, 4, 100 (2021).
[39] Guojing Hu et al. Tunable zero-field superconducting diode effect in two-dimensional ferro-magnetic/superconducting $Fe_3GeTe_2/NbSe_2$ heterostructure. *Nano Research*, 18, 94907068 (2025).
[40] Jiajun Ma, Ruiya Zhan, and Xiao Lin. Superconducting diode effects: Mechanisms, materials and applications. *Advanced Physics Research*, 4, 2400180 (2025).
[41] Zhen Bi, Noah F. Q. Yuan, and Liang Fu. Designing flat bands by strain. *Physical Review B*, 100, 035448 (2019).
[42] Shiyong Tan et al. Interface-induced superconductivity and strain-dependent spin density waves in $FeSe/SrTiO_3$ thin films. *Nature Materials*, 12, 634–640 (2013).
[43] Yuan Hou et al. Strain engineering of twisted bilayer graphene: The rise of strain-twistronics. *Small*, 21, 2311185 (2025).
[44] Asish K. Kundu et al. Charge density waves and the effects of uniaxial strain on the electronic structure of 2H-$NbSe_2$. *Communications Materials*, 5, 208 (2024).
[45] Shang Gao et al. Atomic-scale strain manipulation of a charge density wave. *Proceedings of the National Academy of Sciences*, 115, 6986–6990 (2018).
[46] Paul Dreher et al. Proximity effects on the charge density wave order and superconductivity in single-layer $NbSe_2$. *ACS Nano*, 15, 19430–19438 (2021).
[47] Rui Lou et al. Suppression of nematicity by tensile strain in multilayer $FeSe/SrTiO_3$ films. *Physical Review Research*, 5, 043011 (2023).
[48] Jianyu Xie et al. Fragile pressure-induced magnetism in FeSe superconductors with a thickness reduction. *Nano Letters*, 21, 9310–9317 (2021).
[49] Sha Han, Can-Li Song, Xu-Cun Ma, and Qi-Kun Xue. Interface enhanced superconductivity in $FeSe/SrTiO_3$ and the hidden nature. *Comptes Rendus Physique*, 22, 163–182 (2021).
[50] Satoru Ichinokura, Katsuaki Sugawara, Akari Takayama, Takashi Takahashi, and Shuji Hasegawa. Superconducting calcium-intercalated bilayer graphene. *ACS Nano*, 10, 2761–2765 (2016).

[51] Haruko Toyama et al. Two-dimensional superconductivity of Ca-intercalated graphene on Sic: Vital role of the interface between monolayer graphene and the substrate. *ACS Nano*, 16, 3582–3592 (2022).
[52] Lan Nguyen et al. Atomic defects and doping of monolayer $NbSe_2$. *ACS Nano*, 11, 2894–2904 (2017).
[53] Liangguang Jia et al. Colossal structural distortion and interlayer-coupling suppression in a van der Waals crystal induced by atomic vacancies. *Nano Research*, 16, 5715–5720 (2023).
[54] Guanghui Cheng et al. Reversible modulation of superconductivity in thin-film $NbSe_2$ via plasmon coupling. *Nature Communications*, 15, 6037 (2024).
[55] Claudio Giannetti et al. Ultrafast optical spectroscopy of strongly correlated materials and high-temperature superconductors: A non-equilibrium approach. *Advances in Physics*, 65, 58–238 (2016).
[56] Zhuoyu Chen et al. Carrier density and disorder tuned superconductor-metal transition in a two-dimensional electron system. *Nature Communications*, 9, 4008 (2018).
[57] Rusen Yan et al. Thickness dependence of superconductivity in ultrathin NbS_2. *Applied Physics Express*, 12, 023008 (2019).
[58] Aharon Kapitulnik, Steven A. Kivelson, and Boris Spivak. Colloquium: Anomalous metals: Failed superconductors. *Reviews of Modern Physics*, 91, 011002 (2019).
[59] Yu Saito, Yuichi Kasahara, Jianting Ye, Yoshihiro Iwasa, and Tsutomu Nojima. Metallic ground state in an ion-gated two-dimensional superconductor. *Science*, 350, 409–413 (2015).
[60] Koichiro Ienaga et al. Broadened quantum critical ground state in a disordered superconducting thin film. *Nature Communications*, 15, 2388 (2024).
[61] V. Humbert et al. Overactivated transport in the localized phase of the superconductor-insulator transition. *Nature Communications*, 12, 6733 (2021).
[62] Yukio Tanaka, Shun Tamura, and Jorge Cayao. Theory of Majorana zero modes in unconventional superconductors. *Progress of Theoretical and Experimental Physics*, 2024, 08C105 (2024).
[63] Yuval Oreg, Gil Refael, and Felix von Oppen. Helical liquids and Majorana bound states in quantum wires. *Physical Review Letters.* 105, 177002 (2010).
[64] Kuei-Lin Chiu et al. Integration of graphene-based superconducting quantum circuits in a three-dimensional cavity. *Physical Review Applied*, 23, 034059 (2025).

6 Low-Dimensional Magnetic Materials

Xinchun Du, Jie Li, and Jianxin Zhong

6.1 BASIC OVERVIEW

Low-dimensional magnetic materials encompass a broad class of systems in which magnetic constituents are confined in one or more spatial dimensions, including zero-dimensional (0D) quantum dots, one-dimensional (1D) nanoribbons, two-dimensional (2D) nanosheets, and thin-film structures. Such dimensional confinement significantly modifies the balance among exchange interactions, quantum fluctuations, and surface or interface effects, leading to magnetic behaviors that differ markedly from those of their three-dimensional (3D) counterparts. In particular, reduced dimensionality often enhances fluctuation effects and increases the role of boundaries and interfaces, giving rise to a wide variety of magnetic ordering phenomena and emergent critical behavior. Examples include the potential existence of long-range magnetic order in 2D materials,[1] the tunability of chiral magnetic order in 1D systems,[2] and the macroscopic quantum tunneling effect in 0D materials.[3]

According to the Mermin–Wagner theorem, an ideal 2D Heisenberg model with continuous symmetry cannot sustain stable long-range magnetic order at finite temperatures, as strong thermal fluctuations will disrupt the ordered state.[4] However, in practical low-dimensional materials, intrinsic properties such as anisotropy, interlayer coupling, and exchange interactions can effectively stabilize topological magnetic structures and single-molecule spin states. Therefore, the study of low-dimensional magnetic systems is not only significant for advancing fundamental theories in condensed matter physics but also lays the material foundation for the development of next-generation information technology devices.

6.2 MAIN TYPES OF LOW-DIMENSIONAL MAGNETIC MATERIALS

After outlining the fundamental physical characteristics of low-dimensional systems, this section focuses on their diverse magnetic ordering states. Low-dimensional confinement significantly modulates the interactions between electronic spins, thereby inducing magnetic behaviors distinct from those observed in 3D bulk materials. Several typical low-dimensional magnetic ordering states are systematically discussed below: low-dimensional ferromagnetic materials with a net magnetic moment; low-dimensional antiferromagnetic materials featuring antiparallel spin arrangement and a zero net magnetic moment; and low-dimensional paramagnets that exhibit only a weak response under an external magnetic field. In addition, given the close

 DOI: 10.1201/9781003655305-8

correlation between ferroelectricity and magnetism, this section also addresses low-dimensional ferroelectric and antiferroelectric materials, which are characterized by spontaneous electric polarization and often couple with magnetic order to form multifunctional low-dimensional systems. Through the analysis of these primary categories, a foundation can be established for understanding the physical essence and application potential of low-dimensional magnetic materials.

6.2.1 Low-Dimensional Magnetic Materials

Classified by dimensionality, 0D ferromagnetic materials are confined in all three spatial dimensions, and their physical properties are dominated by 3D quantum confinement effects. Typical representatives include nanoparticles of Fe, Co, Ni, and their alloys, as well as single-molecule magnets. When conventional ferromagnets are downsized to the nanoscale and approach the single-domain critical size, the magnetization reversal mechanism transitions from domain wall displacement to magnetic moment rotation,[5] resulting in the formation of stable single-domain structures accompanied by significantly enhanced coercivity and magnetic anisotropy. In addition, the atomic magnetic moments within nanoparticles exhibit highly ordered arrangements and a macroscopic net magnetic moment, rendering them often superparamagnetic at room temperature while still maintaining strong ferromagnetic order instantaneously. Meanwhile, single-molecule magnets represent an important class of 0D magnetic systems at the molecular scale, characterized by their intrinsic high-spin ground state and significant axial magnetic anisotropy. These features lead to extremely slow magnetization relaxation, allowing the observation of macroscopic hysteresis loops at low temperatures.[6] Furthermore, in certain molecular crystals, discrete molecular units individually exhibit single-molecule magnet behavior and, through weak inter-molecular interactions, cooperatively form long-range ferromagnetic order at extremely low temperatures. Consequently, 0D ferromagnetic systems not only serve as fundamental models for investigating the competition between magnetic order and thermal fluctuations at reduced dimensions, but their controllable quantum states also provide potential physical platforms for high-density information storage and quantum bits.

Furthermore, 1D ferromagnetic materials refer to ferromagnetic systems that are confined to the nanometer or atomic scale in two spatial dimensions while extending in only one dimension. Due to strong quantum confinement effects and an extremely high surface-to-volume ratio, such materials often exhibit magnetic properties distinct from their 3D bulk counterparts, such as significantly enhanced magnetic anisotropy, size-dependent Curie temperature, and unique domain structures. To date, a variety of 1D ferromagnetic materials have been successfully synthesized. For example, Co, Ni, and their alloys (e.g., FeNi and CoPt) nanowires fabricated via electrochemical deposition or vapor-phase methods can maintain stable ferromagnetism at room temperature with high coercivity.[7] In the case of 1D van der Waals materials, CrX_3 (X = Cl, Br, I) nanowires obtained through chemical vapor or liquid-phase exfoliation retain intralayer ferromagnetic coupling in their 1D form, although the Curie temperature is significantly influenced by size and edge states.[8] Furthermore, 1D chain structures designed based on molecular magnets or metal-organic frameworks are formed by bridging magnetic metal ions (e.g., Mn, Fe, and Co) with organic ligands, allowing ferromagnetic exchange interactions to be tuned through chemical design.[9] The ferromagnetism in such materials typically originates from direct or indirect exchange interactions. The distribution of the demagnetizing field and the evolution of the electronic band structure, both arising from their 1D geometry, further govern the distinct magnetization reversal mechanisms and dynamic behaviors.[10]

The experimental discovery of 2D ferromagnetic materials has transcended the understanding established by the Mermin–Wagner theorem, which suggested that long-range magnetic order is difficult to maintain in low-dimensional systems, thereby pushing the study of ferromagnetism to the limit of atomic-layer thickness. These materials not only exhibit novel physical properties such as layer-dependent magnetic coupling and tunable magnetic anisotropy, but also, owing to their compatibility with 2D semiconductor technology, provide an ideal materials platform for constructing low-power spintronic devices. Consequently, they have emerged as a frontier research direction in condensed matter physics and information device technologies. Among these materials, the 2D ferromagnetic semiconductor CrI_3 stands out as one of the most representative.[1, 11–13] CrI_3 exhibits intrinsic ferromagnetism in both monolayer and bulk forms, with Curie temperatures of approximately 45 and 61 K, bandgaps of 1.53 and 1.286 eV, and a saturation magnetic moment of approximately 3.1 μB per Cr^{3+} ion. The magnetic anisotropy of monolayer CrI_3 originates primarily from anisotropic exchange interaction ($\lambda = 0.09$ meV) rather than single-ion anisotropy ($D \approx 0$), with the easy magnetization axis oriented out-of-plane and a total magnetic anisotropy energy of 0.65 meV. Notably, CrI_3 undergoes a crystallographic phase transition near 200 K, with a monoclinic structure in the high-temperature phase and a rhombohedral structure in the low-temperature phase, demonstrating a strong coupling between its magnetic and crystal structures: an anomalous interlayer spacing occurs at the Curie temperature, and an anomaly in magnetic susceptibility is observed at the structural phase transition. Theoretical calculations further predict that the ferromagnetic order in the bulk may persist down to the monolayer, establishing CrI_3 as a key material system for investigating 2D ferromagnetism.

Thin-film ferromagnetic materials represent a significant research direction at the intersection of condensed matter physics and materials science. Their magnetic ordering behavior exhibits rich physical properties that are distinct from those of bulk materials, arising from low-dimensional confinement, interfacial coupling, and strain modulation. As a typical ferromagnetic semiconductor, $Cr_2Ge_2Te_6$ serves as an important material platform for exploring magnetic characteristics at the 2D limit.[14–17] Bulk $Cr_2Ge_2Te_6$ has a Curie temperature of approximately 61–68 K. When thinned down to a monolayer, the Curie temperature decreases significantly to 21 K due to enhanced spin fluctuations resulting from reduced dimensionality, yet intrinsic ferromagnetic order is maintained at low temperatures. The magnetism of this material primarily originates from the $3d^3$ electron configuration of Cr^{3+} ions, with each Cr atom contributing a magnetic moment of approximately 2.8 μB. Regarding magnetic anisotropy, bulk $Cr_2Ge_2Te_6$ exhibits in-plane easy magnetization characteristics. Theoretical studies indicate that at the monolayer limit, its magnetic anisotropy may be significantly enhanced and could even transition to an out-of-plane easy axis. This property plays a crucial role in stabilizing 2D ferromagnetism. Additionally, $Cr_2Ge_2Te_6$

possesses a moderate semiconductor bandgap (0.96 eV, Heyd–Scuseria–Ernzerhof hybrid functional [2006]) and exhibits interlayer ferromagnetic coupling. Owing to the van der Waals forces between layers, thin-layer samples can be prepared via mechanical exfoliation, and effective control of magnetic order can be achieved through means such as gate voltage. These comprehensive characteristics render it highly promising for applications in low-dimensional spintronic devices.

6.2.2 Low-Dimensional Antiferromagnetic Materials

At the 0D scale, research focuses on the stability of antiferromagnetic order, the size dependence of the Néel temperature, and the quantum characteristics of the response to external fields. For example, Wolfgang Wernsdorfer et al.[18] investigated a dimer system consisting of two Mn_4 single-molecule magnets coupled via antiferromagnetic exchange. Through hysteresis loop and quantum tunneling measurements, they found that this coupling leads to a shift in quantum tunneling resonances and the emergence of entangled states during the tunneling process. This result directly demonstrates that the dimer behaves as a quantum-mechanically coupled entity, rather than a simple superposition of two independent units, highlighting the significant influence of quantum effects on antiferromagnetic order in 0D systems. Additionally, in the context of inorganic nanoparticle systems, E. Winkler et al.[19] studied the magnetic behavior of NiO nanoparticles and revealed a structure comprising an "antiferromagnetically ordered core" and a "magnetically disordered surface shell": the core magnetic moments gradually lose the ability to flip freely at relatively higher temperatures (approximately 70 K), while the surface spins freeze in a spin-glass-like manner at lower temperatures (approximately 17 K), with relaxation times following a power-law divergence. By comparing powder samples (with strong interparticle interactions) with polymer-dispersed samples (with weak interactions), the study showed that interparticle interactions increase the core blocking temperature, reduce the overall magnetization, and enhance the frustration of surface spin freezing.

Experimental studies on 1D antiferromagnetic materials have been primarily conducted on several model systems. In $CuGeO_3$,[20] a spin-Peierls transition induced by spin-lattice coupling was observed, confirming the interplay between lattice instability and magnetic order in 1D systems. In $CsNiCl_3$,[21] investigations of integer-spin ($S = 1$) chains revealed the Haldane gap arising from topological origin, clarifying the critical role of quantum numbers in determining the ground state. In near-ideal Heisenberg antiferromagnetic chains such as $SrCuO_2$,[22] experiments provided direct evidence of spin-charge separation in the electronic excitation spectrum: a high-energy dispersion feature around 1.2 eV corresponds to hole excitations, while a low-energy dispersion around 0.3 eV corresponds to spinon excitations. Collectively, these studies demonstrate that strong quantum fluctuations and dimensional confinement in 1D antiferromagnetic chains can stabilize a series of quantum ground states beyond mean-field theory, including spin-Peierls states, topological states, and spin-charge separation, thereby deepening the understanding of low-dimensional quantum magnetism.

Notably, 2D antiferromagnetic materials feature antiparallel alignment of neighboring spins due to negative exchange interactions, resulting in zero net macroscopic magnetization. These materials exhibit high sensitivity to external stimuli and ultra-fast response dynamics, holding significant application potential in spintronics, high-density information storage, and sensing. α-$RuCl_3$ is a 2D layered Mott insulator[23–28] with a Néel temperature ranging from 7 to 14 K; at low temperatures, it undergoes a phase transition into a long-range zigzag antiferromagnetic state. Strong spin–orbit coupling gives rise to local magnetic moments with an effective spin of Jeff = 1/2 on Ru^{3+} ions within the honeycomb lattice, with a magnitude of approximately 0.4 μB, significantly lower than the pure spin value, reflecting the renormalization effect of strong spin–orbit coupling. This material exhibits pronounced in-plane magnetic anisotropy. When an external magnetic field is applied to suppress the zigzag magnetic order, α-$RuCl_3$ can transition into a candidate quantum spin liquid state, wherein half-integer quantized thermal conductivity suggests the presence of chiral Majorana edge modes. Furthermore, α-$RuCl_3$ displays significant spin-phonon coupling and a magnetic-field-induced continuum response in the terahertz frequency range, providing crucial clues for understanding its unconventional excitations.

Notably, 2D thin-film antiferromagnetic materials exhibit strong in-plane spin correlations, along with novel tuning degrees of freedom introduced by dimensionality and interfacial effects. Taking $MnBi_2Te_4$ as an example, this intrinsic magnetic topological insulator has attracted widespread attention in condensed matter physics due to its unique physical properties.[29–34] Bulk $MnBi_2Te_4$ crystals exhibit A-type antiferromagnetic order with a Néel temperature of approximately 24–25 K. When thinned down to the few-layer limit, the material may exhibit ferromagnetic order, with a Curie temperature generally lower than the bulk Néel temperature and a specific value that varies with layer number. Meanwhile, each Mn^{2+} ion located at the center of a [$MnTe_6$] octahedron contributes a high-spin magnetic moment of approximately 5 μB, forming the basis for the strong localized magnetism of the system. Furthermore, this material possesses a significant out-of-plane easy magnetization axis, meaning its magnetic anisotropy is perpendicular to the van der Waals layer plane-a key factor for realizing topological states such as the quantum anomalous Hall effect. More importantly, the most essential property of $MnBi_2Te_4$ lies in the close coupling between its magnetic order and topological states: the antiferromagnetic order corresponds to an axion insulator state. In contrast, the ferromagnetic order corresponds to a quantum anomalous Hall state with Chern number $C = 1$. These rich physical characteristics establish $MnBi_2Te_4$ as an ideal platform for exploring the interplay between magnetism and topological states of matter.

6.2.3 Low-Dimensional Paramagnetic Materials

Notably, 0D paramagnetic materials typically exist in the form of discrete nanoclusters, molecules, or nanoparticles, and their magnetic properties are governed by the discrete energy level distribution of unpaired electrons within confined potentials. Compared with bulk materials, 0D paramagnetism exhibits unique size- and surface-dependent characteristics: from one point of view, quantum size effects lead to energy level discretization, significantly influencing the excitation and relaxation behaviors of local magnetic moments. Taking molecular magnets as an example, single-molecule magnets based on transition metals (such as Mn_{12} and Fe_8) behave as typical 0D paramagnets at high temperatures, with their magnetization behavior describable by the paramagnetic center model. At the same time, the strong intramolecular exchange interactions and magnetic anisotropy lay the foundation for their slow magnetic relaxation at low temperatures.[35]

Notably, 1D ferromagnetic chain materials typically consist of magnetic ions forming strongly ferromagnetically coupled chains, yet exhibit extremely low 3D magnetic ordering temperatures, thereby displaying paramagnetic behavior over a wide temperature range: strong intrachain correlations coexist with weak interchain coupling. For instance, I. Cabrera et al.[36] reported neutron scattering studies of magnetic excitations in the quantum paramagnetic phase of $CoNb_2O_6$ under a transverse magnetic field. The results show that above the critical field, the system exhibits spin-flip quasiparticle excitations, with the strongest dispersion along the chain direction. By parametrically fitting the excitation dispersion using a spin-flip hopping model and linear spin–wave theory, it was found that the experimentally measured bandwidth along the chain direction is significantly smaller than that predicted by spin–wave theory based on zero-field parameters. This deviation is attributed to renormalization effects induced by quantum fluctuations near the critical field, revealing the limitations of classical spin–wave descriptions and the crucial role of quantum fluctuations in strongly correlated 1D systems.

In 2D paramagnets, the spins are typically subject to stronger quantum fluctuations and frustration effects, which may give rise to novel states of matter such as quantum spin liquids and topological order. For instance, in semiconductor quantum wells,[37] the introduction of magnetic impurities generates exchange fields and spin–orbit coupling, enabling the Berry curvature-driven anomalous Hall effect even in the paramagnetic state. Studies show that under specific conditions-such as strong spin–orbit coupling, moderate exchange fields, and low temperatures-this conductance can approach the half-quantized value of $e^2/(2h)$, indicating that 2D paramagnets extend far beyond classical paramagnetic systems, encompassing a research domain where topology, entanglement, and disorder interplay. This holds potential significance for the future development of spintronics and quantum information materials.

Thin-film paramagnets often exhibit novel magnetic states under the influence of 2D confinement and interfacial effects, and have thus become a research focus in the field of low-dimensional magnetism. For instance, the observation of the paramagnetic Meissner effect in superconducting thin films[38] reveals exotic magnetic behavior in low-dimensional systems. In the field of transition metal-doped dilute magnetic semiconductors, Qingyu Xu et al.[39] systematically investigated the magnetic and transport properties of ZnO thin films with varying Co doping concentrations. All samples exhibited only paramagnetism at low temperatures, and X-ray magnetic circular dichroism spectroscopy further confirmed the paramagnetic nature of Co^{2+} ions, ruling out the presence of intrinsic ferromagnetic order. Analysis via Brillouin function fitting revealed that the average magnetic moment of Co^{2+} ions is lower than the theoretical expectation, which is primarily attributed to antiferromagnetic exchange interactions between adjacent Co^{2+} ions and the non-uniform distribution of Co^{2+} in the ZnO matrix, leading to a tendency toward cluster formation. Thin-film paramagnetic systems not only provide an ideal platform for exploring the effects of dimensionality reduction on magnetic interactions but also connect to frontier physics issues such as spin transport, topological effects, and many-body entanglement.

6.2.4 Low-Dimensional Ferroelectric and Antiferroelectric Materials

Confining ferroelectricity to low-dimensional systems not only challenges the physical limits of such materials but also gives rise to novel quantum phenomena. α-In_2Te_3 exhibits both in-plane and out-of-plane ferroelectricity at atomic-layer thickness. This intrinsic 2D ferroelectricity originates from its non-centrosymmetric crystal structure. At the monolayer limit, charge screening effects are sufficient to stabilize the dipole ordering,[40] providing a new pathway for realizing electric-field-tunable topological phase transitions and switching of the quantum anomalous Hall effect. In $MnBi_2Te_4/In_2Te_3$ heterostructures, Jie Li et al.[41] proposed and experimentally validated a novel mechanism for controlling the layer-polarized anomalous Hall effect in even-layer $MnBi_2Te_4$ by balancing the work function difference and the ferroelectric electrostatic potential difference. When the ferroelectric polarization is oriented upward, driven by a net electrostatic potential difference of 0.96 eV, approximately 0.02e of charge is injected from the In_2Te_3 layer into $MnBi_2Te_4$, transforming the interlayer exchange parameter J_Z from −0.010 meV to 0.009 meV. This induces a phase transition from antiferromagnetism to ferromagnetism and gives rise to a nonzero quantum anomalous Hall conductance near the Fermi level, offering a directly measurable transport signal for the electrical control of topological states. This mechanism represents a core physical foundation for realizing low-power spintronic functionalities.

In contrast to ferroelectrics, antiferroelectric materials exhibit antiparallel alignment of adjacent electric dipoles, resulting in nearly zero macroscopic net polarization, yet they feature distinctive double hysteresis loops and high energy storage density. When the classical antiferroelectric material $PbZrO_3$ is deposited as a thin film on a p-type silicon substrate, its macroscopic electric polarization behavior at room temperature exhibits a significant thickness dependence. When the film thickness decreases below a critical value, the antiferroelectric state—characterized by double hysteresis loops in the bulk—is replaced by a ferroelectric state with a single hysteresis loop and pronounced remnant polarization. As the thickness increases and grain size grows, the formation of ferroelectric domains induces depolarization, restoring the bulk antiferroelectric characteristics above a critical thickness.[42] These findings confirm that the coupling between thin-film structure and interfacial electric fields can effectively stabilize the ferroelectric phase within antiferroelectric systems, offering new insights for the design of low-dimensional antiferroelectric functional devices.

This subsection systematically discusses the main types and characteristics of magnetic and dielectric behaviors in low-dimensional systems. The findings indicate that effects such as dimensionality reduction, interfacial coupling, and quantum confinement have a significant impact on the intrinsic parameters and excitation spectra of materials: quantum fluctuations and frustration effects are notably enhanced in low-dimensional ferromagnetic materials; low-dimensional antiferromagnetic systems often exhibit strong spin correlations; low-dimensional paramagnets can give rise to topological transport phenomena such as the quantum anomalous Hall effect through the coupling between charge carriers and local magnetic moments; and in low-dimensional ferroelectric/antiferroelectric materials, size effects and interfacial electric fields can induce polarization ordering. These phenomena reveal the rich potential of low-dimensional materials in correlated electron physics, topological states, and multifunctional devices, while providing key physical insights for understanding the "order–structure–property" relationships under dimensional modulation.

6.3 REGULATION OF LOW-DIMENSIONAL MAGNETIC MATERIALS

Effectively modulating low-dimensional magnetic materials is key to advancing their practical applications. In recent years, various physical and engineering approaches have been developed to precisely control properties such as magnetic order, magnetic anisotropy, and spin dynamics in these materials. These methods primarily include: spin manipulation via external fields; interface-induced magnetic modulation using van der Waals heterostructures; electrostatic fields applied through gate voltages or ionic liquids; strain engineering based on substrates or flexible devices; and synergistic control through other physical fields such as optical excitation, chemical doping, and temperature fields. Summarizing the physical mechanisms, experimental progress, and interrelationships of these modulation strategies is of great significance for deeply understanding the physical origin of low-dimensional magnetism, designing novel spintronic devices, and promoting their functional integration.

6.3.1 Spin Manipulation

Spin manipulation refers to the direct modification of the arrangement, orientation, and dynamic processes of spins in low-dimensional magnetic materials using external fields (such as magnetic fields, optical fields, electric currents, or electric fields). It represents a core approach for achieving magnetic control. Across materials of different dimensionalities, this modulation method exhibits diverse realizations.

In 0D systems, using Fe_3O_4 magnetic nanoparticles as an example, an external magnetic field enables reversible control over their superparamagnetic state-a property that has been widely applied in magnetic targeting and hyperthermia. Subsequently, Pankhurst et al.[43] systematically reviewed the principles and research progress of magnetic fields in nanoparticle assembly, magnetization response, and their biomedical modulation. In 1D materials, Masamitsu Hayashi et al.[44] employed nanosecond spin-polarized current pulses to achieve controlled writing, directional displacement, and real-time detection of magnetic domain walls in Permalloy nanowires, constructing a current-driven three-bit shift register prototype and experimentally validating the operational mechanism of racetrack memory based on the spin-transfer torque effect. In the study of 2D intrinsic magnetic systems, Huang et al.[1] first observed out-of-plane Ising-type ferromagnetic order in monolayer CrI_3 and found that the magnetic order exhibits a pronounced layer dependence: the monolayer exhibits a ferromagnetic state, the bilayer shows interlayer antiferromagnetic coupling, and trilayer and thicker flakes revert to the ferromagnetic state. Moreover, the antiferromagnetic ground state of bilayer CrI_3 is highly sensitive to an external magnetic field, requiring only approximately 0.65 T of out-of-plane field to induce a magnetic phase transition and achieve reversible switching from the antiferromagnetic to the ferromagnetic state, thereby enabling effective control of interlayer spin orientation. In the Fe_3GaTe_2 thin-film system, Shivam N. Kajale et al.[45] combined Fe_3GaTe_2 thin films with a Pt layer possessing strong spin–orbit coupling, demonstrating that the perpendicular spin current generated through the Pt layer can efficiently flip the magnetization direction of Fe_3GaTe_2. This work provides key proof-of-principle validation for the development of low-power, integrable spintronic memory (magnetoresistive random-access memory [MRAM]) based on 2D magnetic materials.

6.3.2 Interface Magnetism Modulation

Interfacial modulation in low-dimensional magnetic materials refers to the control of key magnetic properties—such as magnetic order, anisotropy, and spin dynamics—through the design of chemical, structural, or electronic coupling at interfaces. Its underlying principle is based on interfacial effects (e.g., charge transfer, orbital reconstruction, strain transfer, or proximity coupling) that modify local magnetic exchange interactions and magnetocrystalline anisotropy, thereby tuning the macroscopic magnetic properties of the material.

In 0D magnetic material systems, Tuomo Tanttu et al.[46] systematically investigated silicon metal-oxide-semiconductor double quantum dots located at the Si/SiO_2 interface, revealing the mechanism of spin–orbit interaction modulated by the direction of an external magnetic field, where the spin–orbit coupling is governed by the asymmetry between the interface and the bulk structure. By measuring physical quantities such as the g-factor, the derivative of the Stark shift, the decoherence time T_2, and singlet-triplet mixing, they achieved quantitative characterization and active tuning of the spin–orbit coupling strength. Their results demonstrate that by leveraging the spin–orbit coupling effect, the sensitivity of the Stark shift to charge noise can be effectively suppressed through tuning the external magnetic field direction, leading to an 80% increase in T_2. Furthermore, the magnetic field direction also enables significant control over the probability of spin-flip tunneling between the double quantum dots. G. Gubbiotti et al.[47] systematically investigated the reconfigurability of spin–wave band structures and its underlying physical mechanisms in 1D arrays of Permalloy/Cu/Permalloy (Py/Cu/Py) nanowires. By varying the inter-wire spacing from 80 to 280 nm, they explored the influence of the magnetic field on the switching behavior of the magnetization in the top and bottom Py layers between parallel and antiparallel configurations, revealing the significant modulation of the spin–wave band structure by magnetic configuration switching. The study found that in the antiparallel configuration, the oppositely oriented magnetizations of the two layers result in only one component of the dynamic magnetization precession remaining in phase, thereby weakening the interlayer dynamic dipolar coupling and reducing the bandwidth of tunable spin–wave modes. Only when the precession in the two layers is in a symmetric in-phase mode can spin waves propagate effectively through the array as Bloch waves.

In the field of 2D materials, Ding Zhong et al.[48] constructed WSe_2/CrI_3 heterostructures. Their study revealed that the huge magnetic exchange field of nearly 13 T generated at this heterointerface not only enables effective control of the spin and valley degrees of freedom in monolayer WSe_2 but also allows manipulation of the valley splitting and polarization in WSe_2 by flipping the magnetization direction of CrI_3. Moreover, the photoluminescence intensity of the heterostructure depends on the relative orientation between the photoexcited spins in WSe_2 and the magnetization of CrI_3. This structure not only provides a sensitive local probe for investigating the complex domain dynamics of ultrathin magnets but also offers a powerful platform for advancing the development of spin/valleytronics. L. D. Alegria et al.[49] investigated heterostructures formed by the ferromagnetic insulator $Cr_2Ge_2Te_6$ and the topological insulator Bi_2Te_3. At low temperatures, they found that these heterostructures exhibit a pronounced anomalous Hall effect and hysteretic behavior, indicating that the magnetism of $Cr_2Ge_2Te_6$ is effectively transferred to the Bi_2Te_3 layer via the proximity effect, thereby influencing its transport properties. This discovery provides a promising material platform for topological insulator-based spintronic devices.

6.3.3 Electric Field Modulation

The core mechanisms of electric field modulation in low-dimensional magnetic materials primarily involve two physical effects: voltage-controlled magnetic anisotropy (VCMA) and ionic liquid gating. The VCMA effect modulates magnetic anisotropy by adjusting the density of electronic states at the interface via an electric field, thereby influencing spin–orbit coupling. Ionic liquid gating, conversely, utilizes an electric field to drive the accumulation of ions on the material surface, forming a strong electric double layer that efficiently modifies the carrier concentration through electrostatic doping, thereby inducing magnetic order transitions.

In 0D systems, Faxian Xiu et al.[50] investigated self-assembled $Mn_{0.05}Ge_{0.95}$ dilute magnetic semiconductor quantum dots and successfully achieved electrical control of the ferromagnetism in these quantum dots by applying gate voltage in a metal-oxide-semiconductor capacitor structure. The research team successfully fabricated single-crystalline quantum dots free of metallic precipitates using molecular beam epitaxy, and SQUID measurements confirmed a Curie temperature exceeding 400 K, demonstrating the potential for room-temperature ferromagnetism. Puspamitra Panigrahi et al.[51] performed first-principles calculations to study the magnetic anisotropy of Fe/Pt multilayer nanowires with two different diameters (0.4 and 0.9 nm) and their response to an external electric field. The results show that the non-magnetic Pt spacer layer can significantly modulate the exchange coupling and the direction of the easy magnetization axis of the Fe nanowires, while substantially enhancing the magnetic anisotropy. More importantly, an electric field applied perpendicular to the nanowire axis produces distinctly different effects on the two diameter systems: for the 0.4 nm Fe/Pt nanowire, the electric field has a negligible effect on the magnetic anisotropy, spin moments, and orbital moments; whereas for the 0.9 nm Fe/Pt nanowire, the electric field can effectively modulate the magnetic anisotropy, achieving an enhancement of 31% under an electric field of 0.07 V/Å, accompanied by corresponding changes in the spin and orbital moments of the Fe and Pt atoms.

The research group of Shengwei Jiang[52] reported a breakthrough in electric field control of magnetic order in bilayer CrI_3. By constructing a dual-gate device with the structure graphene/hexagonal boron nitride (G/hBN) /bilayer CrI_3/hBN/G, they observed for the first time a giant linear magnetoelectric effect in a 2D van der Waals magnet. Furthermore, they discovered that in the vicinity of the critical magnetic field for spin flips, a purely electric field can drive fully reversible switching of the magnetic order between antiferromagnetic and ferromagnetic states. Subsequently, the research group of Bevin Huang[53] further reported electric field control of 2D magnetic materials in bilayer CrI_3. By constructing a G/hBN/bilayer CrI_3/hBN/G heterostructure, they realized two types of electrically controlled magnetic behaviors: under an applied magnetic field, the antiferromagnetic-to-ferromagnetic phase transition can be driven by tuning the gate voltage; at zero magnetic field, the net magnetization can be linearly modulated by the gate voltage, revealing an antiferromagnetic state with spin-layer locking. In thin-film materials, Zhi Wang et al.[54] achieved effective control of $Cr_2Ge_2Te_6$ thin films using electrostatic gating techniques. By applying gate voltage to $Cr_2Ge_2Te_6$ field-effect transistors approaching the 2D limit via both ionic liquid gating and silicon back-gating, and performing in situ observations using micro-area magneto-optical Kerr effect measurements below the Curie temperature, they observed reversible modulation of the magnetic hysteresis loops. The study demonstrates that gate-voltage-induced carrier doping can bidirectionally tune the saturation magnetization (M_s) and saturation field (H_s): hole doping enhances M_s while reducing H_s, whereas electron doping produces an even more pronounced enhancement of M_s, with the coercive field remaining essentially unchanged.

6.3.4 Strain Modulation

Strain engineering is a key approach for modulating the properties of low-dimensional magnetic materials. Its underlying principle involves introducing lattice tensile or compressive distortions through the application of external mechanical stress or lattice mismatch in heterostructures, thereby altering atomic bond lengths and bond angles. These changes affect electronic orbital overlap, crystal field symmetry, and the strength of spin–orbit coupling, ultimately enabling the tuning of intrinsic properties such as magnetic exchange interactions, magnetocrystalline anisotropy, and magnetic ordering temperatures.

For 0D systems, E. Zallo et al.[55] utilized strain fields to modulate the coherent tunneling coupling strength of holes in InGaAs/GaAs quantum dot molecules. High-resolution photoluminescence spectroscopy measurements revealed that the energy gap between the bonding and antibonding states in the quantum dot molecules is highly sensitive to strain: tensile strain leads to a linear increase in the coupling strength, while compressive strain reduces it. Eight-band k·p theory calculations indicate that strain directly affects the overlap integral of the wave functions between quantum dots by modifying the effective confinement potentials for heavy holes and light holes in the barrier region. Byron Llerena Zambrano et al.[56] investigated the electrical properties of composite materials under strain by controlling the orientation of 1D nanowires (silver nanowires and gold-coated titanium dioxide nanowires).

The results show that when the nanowires are aligned parallel to the strain direction, the composite exhibits high strain sensitivity with a gauge factor (GF) of up to 80, making it suitable for high-sensitivity strain sensors; when the nanowires are oriented perpendicular to the strain direction, the resistance of the composite shows almost no variation with strain, yielding a GF as low as approximately 0.75.

For 2D materials, Fawei Zheng et al.[57] systematically studied the modulation of the spin state in monolayer CrI_3 by strain and charge doping. By constructing magnetic phase diagrams under the combined effects of biaxial strain and charge doping, they found that strain can effectively tune the magnetic order and magnetic anisotropy: compressive strain (approximately −5.7%) induces a transition from a ferromagnetic insulator to an antiferromagnetic insulator, while tensile strain (approximately 14.3%) can flip the magnetization direction from out-of-plane to in-plane. Additionally, charge doping—particularly electron doping—significantly affects the magnetic anisotropy transition boundary under tensile strain, but has a relatively minor impact on the strain-induced antiferromagnetic-ferromagnetic phase transition. Yanjie Wang et al.[58] employed first-principles calculations to explore the modulation of magnetic anisotropy in monolayer 1T-RuO_2 and 1T-OsO_2 thin films by strain. The results indicate that both systems possess substantial magnetic anisotropy, with monolayer 1T-RuO_2 exhibiting a magnetic anisotropy energy as high as −42.67 meV per unit cell and a Curie temperature above liquid nitrogen temperature. Notably, tensile strain significantly enhances the magnetic anisotropy of monolayer 1T-OsO_2, achieving an increase of approximately 19% under 2% strain. Through density of states and orbital-resolved magnetic anisotropy analyses based on second-order perturbation theory, it was revealed that the large magnetic anisotropy primarily originates from the matrix element differences between the d_{xy} and $d_{x^2-y^2}$ orbitals with opposite spins on the Ru and Os atoms, while the strain-induced enhancement is attributed to the transition of the contributions from the d_{yz} and dz^2 orbitals of the Os atoms from zero to negative values.

6.3.5 Other Regulation Methods

In addition to the above-mentioned modulation methods, key properties of low-dimensional magnetic materials-such as magnetic order, magnetic anisotropy, and Curie temperature-can also be efficiently controlled through multiple physical fields, including optical excitation, chemical doping, and temperature fields. In terms of optical excitation, Nanshu Liu et al.[59] systematically investigated heterostructures composed of group III chalcogenides (MX, where M = In/Ga, X = S/Se) and monolayer CrI_3 using first-principles calculations and nonadiabatic molecular dynamics simulations. The study found that under optical excitation, these heterostructures achieve spin-selective charge separation: holes in the majority spin channel are rapidly injected into the MX layer with a long lifetime, while carriers in the minority spin channel recombine quickly. This spin injection effect, in conjunction with the van Hove singularities in the MX material and interfacial charge transfer, can induce uniform and tunable ferromagnetism in an otherwise non-magnetic MX monolayer, with the magnetic moment and exchange energy varying with the hole doping concentration.

Chemical doping directly modulates local magnetic moments through the introduction of impurities or intercalated ions. For instance, Cuizu Chang et al.[60] first observed the quantum anomalous Hall effect in Cr-doped $(Bi, Sb)_2Te_3$ magnetic topological insulators. This effect achieves quantized Hall resistance (h/e^2) at zero magnetic field, accompanied by a significant drop in longitudinal resistance. The study revealed that in the $Cr_{0.15}(Bi_{0.1}Sb_{0.9})_{1.85}Te_3$ system, the incorporated Cr atoms provide local magnetic moments, enabling the material to spontaneously form ferromagnetic order at low temperatures (approximately 15 K), thereby breaking time-reversal symmetry and opening a magnetic gap in the topological surface states. This discovery laid the foundation for the development of low-power electronic and spintronic devices. As a fundamental modulation approach, the temperature field directly influences magnetic order by altering the strength of thermal fluctuations. B. Huang[1] observed a ferromagnetic phase transition in monolayer CrI_3 at approximately 45 K, slightly lower than that of the bulk material (61 K), indicating weak interlayer coupling. The effect of temperature on magnetic order manifests as follows: below the Curie temperature, the system exhibits ferromagnetic order; as the temperature rises above the Curie point, thermal fluctuations intensify, disrupting the long-range magnetic order and driving the system into a paramagnetic state. Notably, synergistic strategies involving multiple physical fields enable more flexible control. For example, Zhi Wang et al.[54] systematically modulated the magnetic and electronic transport properties of $Cr_2Ge_2Te_6$ thin films by combining electric field, temperature field, and thickness effects. In their experiments, electric fields were applied via both solid-state silicon gates and ionic liquid gates. They found that the ionic liquid gate, due to its electric double layer structure, achieves higher carrier doping concentrations, enabling significant modulation of the saturation field and magnetization of the magnetic hysteresis loop below the Curie temperature (~64 K). Meanwhile, as thickness decreases, the material gradually exhibits insulating behavior, and thin-layer samples encapsulated in hBN still show ambipolar field-effect behavior at low temperatures. Theoretical calculations further indicate that the electric field can shift the Fermi level within the spin-polarized bands, adjusting the relative contributions of Te-p and Cr-d orbitals to the total magnetic moment, thereby achieving electrical modulation of the material's magnetism. These results highlight the critical role of multi-field synergistic modulation in controlling the properties of low-dimensional magnetic materials.

Through a systematic review of the modulation strategies for low-dimensional magnetic materials, it is evident that the field has developed a sophisticated control system characterized by multidimensional and multi-physics-field synergistic approaches. Spin manipulation, which directly acts on magnetic moments via external fields, establishes the foundation for magnetic control. Interfacial modulation leverages heterostructure design, utilizing proximity effects and charge transfer to achieve efficient control of magnetic layers. Electric field modulation alters carrier concentration and Fermi level position through electrostatic doping or ion intercalation, directly adjusting the magnetic ordering state. Strain engineering influences the electronic structure, exchange interactions, and magnetic anisotropy through lattice distortion, providing a viable pathway for stress sensors and flexible magnetoelectric devices. Furthermore, methods such as optical excitation and chemical doping further expand the dimensions of control. In the future, the integration and coupling of multiple modulation approaches will represent a key research direction for achieving multifunctional, intelligent, and device-oriented applications of low-dimensional magnetic materials.

6.4 APPLICATION OF LOW-DIMENSIONAL MAGNETIC MATERIALS

Low-dimensional magnetic materials exhibit significant potential for applications in emerging spintronic devices due to their confined dimensionality and pronounced quantum effects. The core functionality of such devices lies in the manipulation of the spin degree of freedom, including the control of magnetic anisotropy, spin-polarized transport, and magnetic coupling through mechanisms such as interfacial effects, dimensional confinement, and quantum confinement, thereby enabling information storage, transmission, and processing.

Notably, 0D carbon nanomaterials are inherently non-magnetic, but magnetic functionality can be introduced through doping with metal ions such as Fe, Co, Ni, Mn, and Gd.[61] Their magnetism originates from the unpaired electrons of the doped metal ions; for example, the paramagnetism of Gd^{3+} significantly enhances the relaxation rate of water protons, while Fe_3O_4 nanoparticles provide superparamagnetism. The combination of these magnetic functionalities with the intrinsic fluorescence properties of the materials enables key applications in several fields. The applications of 1D magnetic materials mainly focus on two directions: dilute magnetic semiconductors and magnetoresistive devices.[62] In the field of dilute magnetic semiconductors, using ZnO nanowires as an example, doping with transition metal elements such as Mn and Co introduces local magnetic moments into the wide-bandgap semiconductor, achieving ferromagnetic ordering through carrier-mediated indirect exchange interactions. These materials possess relatively high Curie temperatures and are suitable for short-wavelength magneto-optical devices and spin injection devices, providing a material foundation for nanoscale spintronic devices. In the realm of magnetoresistive devices, interfacial effects and antiphase boundaries in core-shell heterostructures (e.g., Fe_3O_4/MgO nanowires) are utilized to modulate spin-dependent transport, achieving a magnetoresistance effect of approximately 1.2% at room temperature. Additionally, nanowire arrays of iron oxides, such as Fe_3O_4, exhibit significant magnetic anisotropy, making them suitable for high-density magnetic storage and magnetic sensors. Overall, the application of 1D magnetic materials centers on tailoring their magnetic properties and charge/spin transport behaviors through dimensional confinement, doping engineering, and heterostructure design, thereby enabling functionalities such as spin-polarized luminescence, magnetoresistive memory, and sensing.

A spin valve is a core spintronic device based on the giant magnetoresistance (GMR) effect, consisting of two ferromagnetic layers separated by a non-magnetic metallic spacer layer. One ferromagnetic layer serves as the "pinned layer" with a fixed magnetization direction, while the other acts as the "free layer" whose magnetization direction varies with an external magnetic field. When the magnetization directions of the two ferromagnetic layers are parallel, electrons with spins aligned in that direction have a higher transport probability, placing the device in a low-resistance state. When the magnetizations are antiparallel, electron scattering at the interface is enhanced, resulting in a high-resistance state. This resistance change enables the conversion of magnetic field signals into electrical signals. In recent years, with the development of 2D magnetic materials, significant progress has been made in spin valve research. The van der Waals spin valve proposed by Cardoso et al.[63] consists of two ferromagnetic insulating layers (CrI_3) sandwiching a 2D non-magnetic conductor (bilayer graphene), with its operating mechanism based on the spin proximity effect. Theoretical calculations and density functional theory simulations both indicate that when the two magnetic layers are parallel, the bilayer graphene remains metallic with high conductance; when they are antiparallel, a bandgap of approximately 7 meV opens at the Dirac point, leading to a significant decrease in conductance, thereby realizing a switching effect based on the magnetic configuration. This type of spin valve offers advantages such as a relatively simple structure, low power consumption, sensitive response, and good compatibility with existing semiconductor processes. However, it generally exhibits a lower magnetoresistance ratio compared to magnetic tunnel junctions (MTJs) and faces challenges such as insufficient high-temperature stability of the pinned layer and a narrow linear operating range. Current research efforts are focused on further enhancing performance and expanding application potential in magnetic sensors and non-volatile memory through the exploration of new material combinations and 3D stacked structures.

An MTJ is a core memory unit in spintronics, typically adopting a sandwich structure consisting of a ferromagnetic layer, an insulating barrier layer, and another ferromagnetic layer. Its operating principle is based on spin-dependent tunneling: when the magnetization directions of the two ferromagnetic layers are parallel, electrons with spins aligned with the layer magnetization have a high tunneling probability, placing the device in a low-resistance state; when the magnetizations are antiparallel, the density of states mismatch suppresses tunneling, resulting in a high-resistance state. This resistance variation, known as the tunneling magnetoresistance (TMR) effect, forms the physical basis for non-volatile information storage and readout. Based on the study by Lin and Chen,[64] a van der Waals MTJ constructed using Fe_3GeTe_2 thin films and hBN (Fe_3GeTe_2/hBN/Fe_3GeTe_2) exhibits excellent transport performance: high conductance is observed in the parallel magnetization state, while conductance is significantly reduced in the antiparallel configuration. Specifically, heterojunctions with monolayer and bilayer Fe_3GeTe_2 achieve TMR ratios of 183 and 252%, respectively, at zero bias, with spin polarization exceeding 75% in the parallel state. The main advantages of MTJs lie in their non-volatility, high read/write speed, virtually unlimited endurance, and compatibility with modern complementary metal–oxide–semiconductor (CMOS) technology. However, challenges remain, including precise control over the thickness and uniformity of the barrier layer, the limitation of high junction resistance in high-speed operation, and the critical influence of interface quality between the ferromagnetic layer and the barrier layer on the tunneling magnetoresistance value. Current research is focused on interface engineering optimization, exploration of novel barrier materials, and the development of 3D integration processes to further enhance device performance and promote large-scale applications in MRAM and spintronic logic chips.

In summary, by virtue of their unique magnetic properties and quantum effects at the nanoscale, low-dimensional magnetic materials provide a critical material foundation for technological innovation across multiple fields. In sensing and energy, they enable high-precision detection and efficient catalysis; in biomedicine, they serve as important platforms for targeted diagnostics and therapies as well as cutting-edge quantum research; in information technology, they drive the miniaturization of ultra-high-density storage devices and low-power spintronic chips. Low-dimensional magnetic materials have become a research focus in materials science and condensed matter physics, acting as a core bridge connecting fundamental physical breakthroughs with next-generation technological applications. They provide robust support for the continued advancement of information technology, energy technology, and quantum technology.

This chapter systematically reviews the research progress and framework of low-dimensional magnetic materials. It begins by introducing the fundamental concepts and development background of such materials. Subsequently, based on magnetic ordering types and functional characteristics, it summarizes and compares the main categories of low-dimensional magnetic materials, including ferromagnetic, antiferromagnetic, paramagnetic, and ferroelectric/antiferroelectric materials. Furthermore, focusing on the key modulation strategies in this field, it elaborates on the mechanisms by which various physical external fields—such as spin manipulation, interfacial magnetic modulation, electric field modulation, and strain engineering—affect the magnetic properties of these materials. Finally, it explores the application potential of such materials in emerging fields, including novel information storage, spintronic devices, and multifunctional integrated systems, aiming to provide researchers with a clear and comprehensive knowledge framework for low-dimensional magnetic materials and to serve as a reference for future related research and technological development.

REFERENCES

[1] Bevin Huang et al., Layer-dependent ferromagnetism in a van der Waals crystal down to the monolayer limit. *Nature* 546, 270–273 (2017).

[2] Y. Togawa et al., Chiral magnetic soliton lattice on a chiral helimagnet. *Phys. Rev. Lett.* 108, 107202 (2012).

[3] Jonathan R. Friedman et al., Macroscopic measurement of resonant magnetization tunneling in high-spin molecules. *Phys. Rev. Lett.* 76, 3830 (1996).

[4] N. D. Mermin and H. Wagner, Absence of ferromagnetism or antiferromagnetism in one- or two-dimensional isotropic heisenberg models. *Phys. Rev. Lett* (1966).

[5] B. D. Cullity and C. D. Graham. *Introduction to Magnetic Materials*. Hoboken, NJ: John Wiley & Sons (2011).

[6] Yuan-Yuan Zhu et al., Cobalt(II) coordination polymer exhibiting single-ion-magnet-type field-induced slow relaxation behavior. *Inorg. Chem.* 54, 3716–3718 (2015).

[7] K. Nielsch et al., Hexagonally ordered 100 nm period nickel nanowire arrays. *Appl. Phys. Lett.* 79, 1360–1362 (2001).

[8] Wei Jiang et al., First-principles calculations of magnetic edge states in zigzag CrI_3 Nanoribbons. *Phys. Lett. A* 383, 754–758 (2018).

[9] Xiangyu Liu et al., One-dimensional cobalt(II) coordination polymer featuring single-ion-magnet-type field-induced slow magnetic relaxation. *New J. Chem.* 42, 9612 (2018).

[10] Akhila Priya Kotti and Amaresh Chandra Mishra, Magnetization reversal and domain wall pinning in nanowires with nonlinear modulation of diameter. *Mater. Lett.* 360, 135990 (2024).
[11] Michael A. McGuire et al., Coupling of crystal structure and magnetism in the layered, ferromagnetic insulator CrI_3. *Chem. Mater.* 27, 612–620 (2015).
[12] Wei-Bing Zhang et al., Robust intrinsic ferromagnetism and half semiconductivity in stable two-dimensional single-layer chromium trihalides. *J. Mater. Chem. C*, 3, 12457 (2015).
[13] J. L. Lado and Joaquín Fernández-Rossier, On the origin of magnetic anisotropy in two dimensional CrI_3. *2D Mater.* 4, 035002 (2017).
[14] Cheng Gong et al., Discovery of intrinsic ferromagnetism in two-dimensional van der Waals crystals. *Nature* 546, 265–269 (2017).
[15] Mohammed Moaied, First-principles prediction of a two-dimensional semiconductor ferromagnetism in $Cr_2X_2Se_6$ (X = Al, Ga, In) materials. *Chin. J. Phys.* 87, 354–363 (2024).
[16] Y. Carteauxt et al., Crystallographic, magnetic and electronic structures of a new layered ferromagnetic compound $Cr_2Ge_2Te_6$. *J. Phys.: Condens. Matter* 7, 69–87 (1995).
[17] Changsheng Song et al., Tunable band gap and enhanced ferromagnetism by surface adsorption in monolayer $Cr_2Ge_2Te_6$. *Phys. Rev. B* 99, 214435 (2019).
[18] Wolfgang Wernsdorfer et al., Exchange-biased quantum tunnelling in a supramolecular dimer of single-molecule magnets. *Nature* 416, 406–409 (2002).
[19] E. Winkler et al., Surface spin-glass freezing in interacting core shell NiO nanoparticles. *Nanotechnology* 1, 9 (2008).
[20] Masashi Hase, Spin-Peierls transition in $CuGeO_3$. *Physica B* 237–238, 123–126 (1997).
[21] W. J. L. Buyers et al., Experimental evidence for the Haldane gap in a Spin-1, nearly isotropic, antiferromagnetic chain. *Phys. Rev. Lett.* 56, 371 (1986).
[22] C. Kim et al., Observation of spin-charge separation in one-dimensional $SrCuO_2$. *Phys. Rev. Lett.* 77, 4054 (1996).
[23] K. W. Plumb et al., α-$RuCl_3$: A spin-orbit assisted Mott insulator on a honeycomb lattice. *Phys. Rev. B*, 90, 041112(R) (2014).
[24] J. A. Sears et al., Magnetic order in α-$RuCl_3$: A honeycomb-lattice quantum magnet with strong spin-orbit coupling. *Phys. Rev. B* 91, 144420 (2015).
[25] Yumi Kubota et al., Successive magnetic phase transitions in α-$RuCl_3$: XY-like frustrated magnet on the honeycomb lattice. *Phys. Rev. B*, 91, 094422 (2015).
[26] A. Banerjee et al., Proximate Kitaev quantum spin liquid behaviour in a honeycomb magnet. *Nat. Mater.* 15, 733–740 (2016).
[27] Y. Kasahara et al., Majorana quantization and half-integer thermal quantum Hall effect in a Kitaev spin liquid. *Nature* 559, 227–231 (2018).
[28] Luke J. Sandilands et al., Spin-orbit excitations and electronic structure of the putative Kitaev magnet α-$RuCl_3$. *Phys. Rev. B* 93, 075144 (2016).
[29] Jiaheng Li et al., Intrinsic magnetic topological insulators in van der Waals layered MnBi2Te4-family materials. *Sci. Adv.* 5, eaaw5685 (2019).
[30] M. M. Otrokov et al., Prediction and observation of an antiferromagnetic topological insulator. *Nature* 576, 416–422 (2019).
[31] Yujun Deng et al., Quantum anomalous Hall effect in intrinsic magnetic topological insulator $MnBi_2Te_4$. *Science*, 367, 895–900 (2020).
[32] Yan Gong et al., Experimental realization of an intrinsic magnetic topological insulator. *Chin. Phys. Lett.* 36, 076801 (2019).
[33] Dongqin Zhang et al., Topological axion states in the magnetic insulator $MnBi_2Te_4$ with the quantized magnetoelectric effect. *Phys. Rev. Lett.* 122, 206401 (2019).

[34] J.-Q. Yan et al., Crystal growth and magnetic structure of MnBi2Te4. *Phys. Rev. Mater.* 3, 064202 (2019).
[35] R. Sessoli et al., Magnetic bistability in ametal-ion cluster. *Nature* 365, 141–143 (1993).
[36] I. Cabrera et al., Excitations in the quantum paramagnetic phase of the quasi-one-dimensional Ising magnet $CoNb_2O_6$ in a transverse field: Geometric frustration and quantum renormalization effects. *Phys. Rev. B*, 90, 014418 (2014).
[37] Dimitrie Culcer, Allan MacDonald, and Qian Niu, Anomalous hall effect in two dimensional paramagnetic systems. *Phys. Rev. B*, 68, 045327 (2003).
[38] D. A. Luzhbin et al., Origin of paramagnetic magnetization in field-cooled $YBa_2Cu_3O_7$-δ film. *Phys. Rev. B*, 69, 024506 (2004).
[39] Qingyu Xu et al., Paramagnetism in Co-doped ZnO films. *J. Phys. D: Appl. Phys.* 42, 085001 (2009).
[40] Yue Li, Ming Gong, and Hualing Zeng, Atomically thin α-In_2Se_3: an emergent two-dimensional room temperature ferroelectric semiconductor. *J. Semicond.* 40(6), 061002 (2019).
[41] Jie Li et al., Ferroelectric-driven magnetic phase transition in topological antiferromagnets. *Phys. Rev. B*, 112, 014418 (2025).
[42] Pushan Ayyub et al., Ferroelectric behavior in thin films of antiferroelectric materials. *Phys. Rev. B* 57, R5559 (1998).
[43] Q. A. Pankhurst et al., Applications of magnetic nanoparticles in biomedicine. *J. Phys. D: Appl. Phys.* 36, R167–R181 (2003).
[44] Masamitsu Hayashi et al., Current-controlled magnetic domain-wall nanowire shift register. *Science* 320, 209–211 (2008).
[45] Shivam N. Kajale et al., Current-induced switching of a van der Waals ferromagnet at room temperature. *Nat. Commun.* 15, 1485 (2024).
[46] Tuomo Tanttu et al., Controlling spin-orbit interactions in silicon quantum dots using magnetic field direction. *Phys. Rev. X*, 9, 021028 (2019).
[47] G. Gubbiotti et al., Interplay between intra- and inter-nanowires dynamic dipolar interactions in the spin wave band structure of Py/Cu/Py nanowires. *Sci. Rep.* 9, 4617 (2019).
[48] Ding Zhong et al., Van der Waals engineering of ferromagnetic semiconductor heterostructures for spin and valleytronics. *Sci. Adv.* 3, e1603113 (2017).
[49] L. D. Alegria et al., Large anomalous Hall effect in ferromagnetic insulator topological insulator heterostructures. *Appl. Phys. Lett.* 105, 053512 (2014).
[50] Faxian Xiu et al., Electric-field-controlled ferromagnetism in high-Curie-temperature Mn0.05Ge0.95 quantum dots. *Nat. Mater.* 9, 337–344 (2010).
[51] Puspamitra Panigrahi and M. C. Valsakumar, Enhanced magnetic anisotropy energy of Fe/Pt nanowires in presence of finite electric field. *IEEE International Conference on Nanoscience, Technology and Societal Implications*. IEEE, 1–4 (2011).
[52] Shengwei Jiang, Jie Shan, and Kin Fai Mak, Electric-field switching of two-dimensional van der Waals magnets. *Nat. Mater.* 17, 406–410 (2018).
[53] Bevin Huang et al., Electrical control of 2D magnetism in bilayer CrI_3. *Nat. Nanotechnol.* 13, 544–548 (2018).
[54] Zhi Wang et al., Electric-field control of magnetism in a few-layered van der Waals ferromagnetic semiconductor. *Nat. Nanotechnol.* 13, 554–559 (2018).
[55] E. Zallo et al., Strain-induced active tuning of the coherent tunneling in quantum dot molecules. *Phys. Rev. B*, 89, 241303(R) (2014).
[56] Byron Llerena Zambrano et al., Magnetic manipulation of nanowires for engineered stretchable electronics. *ACS Nano*, 16, 837–846 (2022).
[57] Fawei Zheng et al., Tunable spin states in the two-dimensional magnet CrI_3. *Nanoscale*, 10, 14298 (2018).

[58] Yanjie Wang et al., Large magnetic anisotropy and its strain modulation in two-dimensional intrinsic ferromagnetic monolayer RuO_2 and OsO_2. *Phys. Chem. Chem. Phys.* 20, 28162 (2018).
[59] Nanshu Liu, Si Zhou, and Jijun Zhao. Photoinduced spin injection and ferromagnetism in 2D Group III monochalcogenides. *J. Phys. Chem. Lett.* 13, 590–597 (2022).
[60] Cui-Zu Chang, Experimental observation of the quantum anomalous hall effect in a magnetic topological insulator. *Science* 167, 1234414 (2013).
[61] Zheng Yang et al., Zero-dimensional carbon nanomaterials for fluorescent sensing and imaging. *Chem. Rev.* 123, 11047–11136 (2023).
[62] Jia Grace Lu, Paichun Chang, and Zhiyong Fan, Quasi-one-dimensional metal oxide materials-Synthesis, properties and applications. *Mater. Sci. Eng. R*, 52, 49–91 (2006).
[63] C. Cardoso et al., Van der Waals spin valves. *Phys. Rev. Lett.*, 121, 067701 (2018).
[64] Zheng-Zhe Lin and Xi Chen, Ultrathin scattering spin filter and magnetic tunnel junction implemented by ferromagnetic 2D van der Waals material. *Adv. Electron. Mater.* 6, 1900968 (2020).

Part III

Cutting-Edge Research and Future Applications

7 Quantum Computing

Junjie Liu and Jianxin Zhong

7.1 QUBITS

The cornerstone of quantum information theory is the quantum bit, or qubit. A qubit is a two-state (or two-level) quantum mechanical system, one of the simplest quantum systems displaying the peculiarity of quantum mechanics. Examples include the spin of the electron, in which the two levels can be taken as spin up and spin down. In a classical system, a bit would have to be in one state or the other. However, quantum mechanics allows the qubit to be in a coherent superposition of multiple states simultaneously, a property that is fundamental to quantum mechanics and quantum computing. Unlike a classical bit, which is restricted to a discrete binary state 0 or 1, a qubit resides in a two-dimensional (2D) complex Hilbert space $\mathcal{H}^2$. Any arbitrary state $|\psi\rangle \in \mathcal{H}^2$ can be expressed as a coherent linear superposition of the computational basis states $\{|0\rangle, |1\rangle\}$ which are eigenstates of the Pauli matrix σ_z,

$$|\psi\rangle = \alpha|0\rangle + \beta|1\rangle \tag{7.1}$$

Here, the complex probability amplitudes α and β satisfy the normalization condition $|\alpha|^2 + |\beta|^2 = 1$.

Geometrically, a qubit is often visualized using the Bloch Sphere. The surface of the Bloch sphere is a 2D space, which represents the observable state space of the pure qubit states. This state space has two local degrees of freedom, which can be represented by the two polar angles θ and ϕ[1]

$$|\psi\rangle = \cos\left(\frac{\theta}{2}\right)|0\rangle + e^{i\phi}\sin\left(\frac{\theta}{2}\right)|1\rangle \tag{7.2}$$

where θ determines the relative weighting (probability) and ϕ defines the relative phase. While a single qubit can represent an infinite number of states, the act of measurement collapses the wavefunction into one of the basis states, yielding only one bit of classical information.

The power of the qubit lies not just in superposition, but in the no-Cloning theorem,[2] which dictates that an unknown quantum state cannot be perfectly copied. This principle provides the fundamental security for quantum communication. Furthermore, computational progress is achieved through quantum gates and unitary transformations U that rotate the state vector on the Bloch Sphere. While a single qubit offers infinite states, it provides only one bit of information upon measurement

DOI: 10.1201/9781003655305-10

due to wavefunction collapse. The true "quantum advantage" emerges from n-qubit systems where the Hilbert space grows as 2^n, allowing for quantum entanglement to link these dimensions in ways classical bits never could. Physical realizations of qubits vary across different platforms. In superconducting circuits, the qubit is typically formed by the nonlinear energy levels of a Josephson junction (e.g., the Transmon qubit). In trapped ion systems, qubits are encoded in the hyperfine ground states or optical transitions of individual ions. Other prominent platforms include semiconductor quantum dots (utilizing electron spin) and photonic systems (utilizing polarization or time-bin encoding).

7.2 PHYSICAL EFFECTS IN QUANTUM COMPUTING

7.2.1 Quantum Entanglement

Quantum entanglement is perhaps the most distinctive feature of quantum mechanics, describing a condition where the collective state of multiple particles cannot be factored into the product states of the individual particles. For a bipartite system, an entangled state $|\Psi\rangle_{AB}$ is one where

$$|\Psi\rangle_{AB} \neq |\psi\rangle_A \otimes |\phi\rangle_B \tag{7.3}$$

This phenomenon represents a form of nonlocal correlation that transcends classical physics. The simplest system to display quantum entanglement is the system of two qubits. For example, the Bell state $|\Phi^+\rangle = \frac{1}{\sqrt{2}}(|00\rangle + |11\rangle)$ implies that if qubit A is measured as $|0\rangle$, qubit B is instantaneously determined to be $|0\rangle$, regardless of the spatial distance between them.

Quantum entanglement also allows multiple states (such as the Bell state mentioned above) to be acted on simultaneously, unlike classical bits that can only have one value at a time. Entanglement is a necessary ingredient of any quantum computation that cannot be done efficiently on a classical computer. Many of the successes of quantum computation and communication, such as quantum teleportation and superdense coding, make use of entanglement, suggesting that entanglement is a resource that is unique to quantum computation. A major hurdle facing quantum computing, as of 2018, in its quest to surpass classical digital computing, is noise in quantum gates that limits the size of quantum circuits that can be executed reliably.

7.2.2 Quantum Coherence and Decoherence Mechanisms

The success of quantum computation hinges on the maintenance of quantum coherence—the stable phase relationship between the states in a superposition. Coherence is represented by the off-diagonal elements of the density matrix ρ. However, no quantum system is perfectly isolated. The interaction between the qubit and its surrounding environment (the "bath") leads to decoherence, effectively causing the system to leak information and revert to a classical probabilistic state. The evolution of such an open quantum system is typically modeled using the Lindblad master equation[3]:

$$\frac{d\rho}{dt} = -i\left[H,\rho\right] + \sum_n \left(L_n \rho L_n^\dagger - \frac{1}{2}\left\{ L_n^\dagger Ln, \rho \right\} \right) \tag{7.4}$$

where $\{L_n\}$ are the Lindblad operators representing different noise channels (bit-flip, phase-flip, etc.). For qubits, there are two primary timescales used to characterize decoherence:

- T_1 (Longitudinal relaxation time): This represents energy relaxation. It is the process by which a qubit in the excited state loses energy to the environment and decays to the ground state.
- T_2 (Transverse dephasing time): This represents the loss of phase information. Even if no energy is exchanged, fluctuations in the environment (e.g., magnetic field noise) cause the relative phase ϕ to drift randomly. T_2 is always less than or equal to $2T_1$.

Minimizing decoherence through cryogenics, vacuum shielding, and material purification is the primary challenge in scaling quantum hardware. To combat decoherence, we must understand the spectral density of the noise. If the noise is "colored" (non-Markovian), we can use dynamical decoupling (as discussed in the following subsection) to effectively filter out the interference. In materials like graphene or diamond, the goal is to engineer a "phononic vacuum" where the coupling between the qubit and the lattice vibrations is minimized, extending the T_1 and T_2 times to the limits required by the threshold theorem for quantum error correction.[4–7]

7.2.3 Quantum Spin–Orbit Coupling

In solid-state quantum computing, specifically within semiconductor quantum dots, spin–orbit coupling (SOC) is a relativistic interaction that links an electron's spin S to its orbital motion p through an effective magnetic field. The Hamiltonian for SOC in a 2D electron gas is often described by the Rashba and Dresselhaus terms:

$$H_{SO} = \alpha\left(\sigma_x p_y - \sigma_y p_x\right) + \beta\left(\sigma_x p_x - \sigma_y p_y\right) \tag{7.5}$$

SOC is a "double-edged sword." On the one hand, it acts as a source of dephasing because spatial fluctuations translate into spin noise. On the other hand, it is an indispensable tool for quantum control. It enables electric dipole spin resonance, allowing researchers to manipulate spin states using local electric fields (via gates) rather than global magnetic fields. This is significantly faster and more scalable. Furthermore, SOC is a key ingredient in generating Majorana bound states in topological superconductors, which are the basis for fault-tolerant topological quantum computing.

7.2.4 Quantum Control

Quantum control is the field of engineering specific time-dependent Hamiltonians $H(t)$ to drive a system from an initial state $|\psi_i\rangle$ to a target state $|\psi_f\rangle$ with high fidelity. The most fundamental control mechanism is Rabi oscillation, where a resonant drive (microwave or laser) causes the qubit state to rotate periodically around the Bloch sphere. Modern quantum control employs several sophisticated techniques.

Pulse Shaping: Instead of simple square pulses, researchers use "derivative removal by adiabatic gate" pulses,[8] that is, standard Gaussian pulses with additional Gaussian derivative components and lifting applied, to prevent leakage into higher, non-computational energy levels.

Dynamical Decoupling[9, 10]: Applying a sequence of π-pulses (like the Hahn Echo) to "refocus" the qubit and cancel out slow-moving environmental noise, effectively extending T_2.

Optimal Control Theory: Algorithms such as gradient ascent pulse engineering[11] are used to find the mathematically optimal pulse shape that maximizes gate fidelity while minimizing time and power.

As we move toward the Noisy Intermediate-Scale Quantum (NISQ) era, the precision of quantum control determines the "depth" of the circuits that can be executed before decoherence renders the result meaningless.

7.3 CURRENT STATUS OF LOW-DIMENSIONAL QUANTUM MATERIALS IN QUANTUM COMPUTING

7.3.1 Graphene and Quantum Computing

Graphene offers high mobility and "valley" degrees of freedom.[12–15] It is the ideal candidate for "flying qubits" and ballistic transport.

7.3.1.1 Unique Electronic Properties

The emergence of graphene as a pivotal candidate for quantum information processing is not merely a result of its high conductivity, but rather stems from the profound symmetry of its honeycomb lattice. This sp^2-hybridized carbon structure gives rise to electronic behaviors that are qualitatively different from those found in traditional three-dimensional semiconductors.[16] In the context of quantum computing, graphene offers a "quiet" environment–one where quantum states can be manipulated with high precision and stored with minimal interference from the host crystal. Particularly, graphene has the following novel features.

At the heart of graphene's physics is its unique energy–momentum relationship. Unlike conventional semiconductors, where the dispersion is parabolic ($E \propto k^2$), electrons in graphene follow a linear, ultra-relativistic-like dispersion near the six corners of the Brillouin zone. These points, known as Dirac points (K and K'), represent where the conduction and valence bands meet. Near the inequivalent Brillouin zone corners, K and K', the tight-binding Hamiltonian reduces to the 2D Dirac form

$$H = \hbar vF\left(\tau_z \sigma_x k_x + \sigma_y k_y\right) \tag{7.6}$$

In this expression, σ denotes the sublattice space, $\tau_z = \pm 1$ serves as the valley index, and $v_F \approx 10^6$ m/s represents the Fermi velocity. This mathematical framework leads to a linear dispersion relation:

$$E(k) = \pm\hbar v_F |k| \tag{7.7}$$

This linear spectrum implies that electrons in graphene behave as massless Dirac fermions,[17–20] possessing a vanishing effective mass and unique chiral eigenstates where the spin (or pseudospin) is locked to the direction of momentum. A significant consequence of this chirality is the suppression of 180° backscattering in the presence of smooth disorder, which facilitates exceptionally high carrier mobility. From a quantum information perspective, these properties enable ballistic transport over micrometer scales, significantly reducing phase randomization caused by impurity scattering. However, because graphene is naturally gapless, achieving the spatial localization and confinement of carriers required for qubits remains a significant engineering challenge, necessitating advanced techniques such as nanostructuring or strain engineering.

Beyond its transport properties, graphene is an exceptional candidate for spin-based quantum memory due to its weak intrinsic SOC. This weakness arises because carbon possesses a low atomic number (Z = 6), and the planar sp^2 symmetry inherently suppresses certain SOC channels. The intrinsic SOC term is expressed as

$$H_{\text{SOC}} = \lambda_{\text{SO}} \tau_z \sigma_z s_z \tag{7.8}$$

where the coupling strength λ_{SO} is on the order of μeV. For a spin qubit encoded in $|\uparrow\rangle, |\downarrow\rangle$ states, the spin relaxation and dephasing rates Γ scale quadratically with the SOC strength: $\Gamma \propto \lambda_{\text{SO}}^2$.

Consequently, the weak SOC leads to a correspondingly weak spin–phonon coupling, which facilitates potentially long T_1 coherence times. From the perspective of open quantum systems, the evolution of the spin density matrix is governed by

$$\frac{d\rho}{dt} = -i[H_s, \rho] + \mathcal{L}_{\text{spin}}(\rho) \tag{7.9}$$

In this system, the dissipator $\mathcal{L}_{\text{spin}}(\rho)$ is significantly suppressed relative to traditional III–V semiconductors, positioning graphene as a highly promising host for robust, spin-based quantum information processing. For a spin qubit, the relaxation time (T_1) and dephasing time (T_2) are inversely proportional to the square of the SOC strength. Consequently, the ultra-weak SOC in graphene creates a "spin-friendly" environment where spin states are largely decoupled from the lattice vibrations (phonons), theoretically allowing for significantly longer coherence times compared to heavier III–V semiconductors like GaAs or InAs.

Another critical factor for spin coherence in graphene is the hyperfine interaction between the electron spin (S) and the surrounding nuclear spin bath ($\{I\}_i$), described by the Hamiltonian:

$$H_{hf} = \sum_{i} A_I S \cdot I_I \tag{7.10}$$

In natural graphene, approximately 98.9% of the material consists of ^{12}C, which possesses a nuclear spin of $I = 0$. This naturally "quiet" environment can be further optimized through isotopic purification to nearly eliminate the nuclear spin bath and its associated magnetic noise. According to the central-spin model, the dephasing rate resulting from nuclear fluctuations scales with the number of spins (N_{spin}) as

$$\frac{1}{T_2^*} \propto \sqrt{N_{\text{spin}}} \tag{7.11}$$

By reducing N_{spin}, the coherence time T_2^* is directly increased, potentially bringing graphene closer to the "nuclear-spin-free" limit seen in isotopically purified silicon.

In addition to charge and spin, graphene possesses a unique binary degree of freedom called the valley index. As the K and K' Dirac points are momentum-separated and energetically equivalent, they can be treated as a pseudospin. While valley degeneracy can complicate certain operations (such as spin-to-charge conversion), it also provides a new paradigm for quantum encoding. Valleytronics explores the use of these valleys as a robust basis for qubits. As the K and K' valleys are separated by a large momentum vector, intervalley scattering requires a short-range potential (like an atomic defect), making the valley index remarkably stable against long-range smooth potentials. This stability offers a layer of protection for quantum information encoded in the valley state.

Graphene is a zero-gap semimetal characterized by a vanishing density of states at the Dirac point. It exhibits perfect ambipolarity, meaning the charge carrier type can be continuously and symmetrically tuned between electrons and holes by an external gate voltage.

In quantum computing architectures, this allows for the "electrostatic painting" of potential landscapes. Researchers can define quantum dots, barriers, and channels within a single graphene sheet by simply modulating local gates. This high degree of tunability is essential for the spatial control of qubits and the implementation of tunable coupling between adjacent qubits, providing a flexible platform for scalable quantum architectures.

7.3.1.2 Graphene Quantum Dots (GQDs)

While the intrinsic electronic properties of graphene provide favorable conditions for long coherence times, such as weak SOC and minimal hyperfine interaction, these advantages alone are insufficient for quantum computation. A scalable quantum architecture requires the ability to localize and individually address single electronic degrees of freedom. In a gapless Dirac material, however, confinement is fundamentally nontrivial.

GQDs, therefore, represent a crucial step in translating graphene's symmetry-protected internal structure into practical qubit platforms. They provide spatial localization of charge carriers, discretization of energy levels, and controlled coupling to external reservoirs–all prerequisites for defining and manipulating two-level quantum systems. Compared to conventional semiconductor-based quantum dots, GQDs have the following unique properties.

In conventional semiconductors, electrostatic gating readily defines quantum dots because electrons possess a finite effective mass and parabolic dispersion. In monolayer graphene, however, low-energy carriers obey a Dirac equation and exhibit Klein tunneling: for normal incidence, transmission through a potential barrier approaches unity. This relativistic-like tunneling prevents purely electrostatic confinement in pristine monolayer graphene. Therefore, the formation of GQDs requires either symmetry breaking or bandgap engineering. In Bernal-stacked bilayer graphene, applying a vertical electric displacement field creates a potential difference between the two layers, opening a tunable energy gap in the otherwise semi-metallic band structure. The low-energy effective Hamiltonian can be expressed as

$$H_{BLG} = \begin{pmatrix} \Delta & \left(\pi^{\dagger}\right)^2 / 2m^* \\ \pi^2 / 2m^* & -\Delta \end{pmatrix} \tag{7.12}$$

where Δ represents the gap induced by the external field and m^* is the effective mass of the electrons. This tunable gap enables electrostatic confinement of charge carriers in a manner analogous to conventional semiconductors.

By adjusting Δ and local gate potentials, one can define quantum dots in bilayer graphene, creating localized electronic states. These confined states provide a practical platform for implementing qubits based on electron spin or the valley degree of freedom, making bilayer graphene a promising material for scalable quantum information processing. Once confinement is achieved, the electronic Hamiltonian reduces to a finite-level interacting system,

$$H = \sum_i \epsilon_i c_i^{\dagger} c_i + \frac{e^2}{2C} \hat{N}^2 \tag{7.13}$$

At temperatures satisfying $k_B T \ll E_C = \frac{e^2}{2C}$, transport through the dot enters the Coulomb blockade regime. The additional energy

$$E_{\text{add}} = E_C + \Delta\epsilon \tag{7.14}$$

includes both the charging and orbital quantization contributions. As graphene systems can exhibit relatively large level spacing $\Delta\epsilon$, few-electron regimes are experimentally accessible, facilitating the isolation of well-defined qubit states.

7.3.1.3 Valleytronics

Valleytronics exploits the valley degree of freedom in graphene as a controllable quantum number for information encoding. In the low-energy description, the two inequivalent Dirac points K and K' define a binary index $\tau = \pm 1$, which can be interpreted as a pseudospin. Unlike real spin, this degree of freedom originates from lattice symmetry and momentum-space topology rather than intrinsic angular momentum.

The effective low-energy Hamiltonian incorporating Berry curvature effects may be written as

$$H_{\text{eff}} = H_0 + \tau z\Omega(\mathrm{k}) \tag{7.15}$$

where $\Omega(k)$ represents the Berry curvature contribution. As time-reversal symmetry enforces opposite Berry curvature in the two valleys, carriers in K and K' respond differently to external fields. This underlies the valley Hall effect, in which an in-plane electric field induces transverse valley currents with opposite polarity.

From a quantum information perspective, the valley subspace forms a natural two-level system (TLS),

$$|0\rangle = |K\rangle, |1\rangle = |K'\rangle \tag{7.16}$$

which may serve as a logical qubit. The robustness of this encoding arises from the large momentum separation between the valleys: intervalley transitions require short-range, atomic-scale disorder. Smooth electrostatic potentials conserve the valley index, suppressing long-wavelength dephasing mechanisms.

Decoherence of a valley qubit can be modeled phenomenologically by an intervalley scattering channel,

$$\mathcal{L}_{\text{valley}}(\rho) = \gamma_v\left(\tau_+\rho\tau_- - \frac{1}{2}\{\tau+, \rho\}\right) \tag{7.17}$$

where γ_v characterizes the intervalley scattering rate. Since γ_v is primarily induced by lattice-scale disorder, valley qubits may exhibit resilience against charge noise that varies smoothly on mesoscopic scales. More advanced architectures exploit the combined spin–valley Hilbert space, allowing the engineering of symmetry-protected subspaces or hybrid qubits with enhanced noise tolerance.

$$H = H_{\text{spin}} \otimes H_{\text{valley}} \tag{7.18}$$

In this sense, valleytronics does not merely introduce an additional degree of freedom but expands the controllable internal structure of graphene-based quantum systems.

7.3.1.4 Moiré Physics and Twisted Bilayer Graphene

A qualitatively new regime emerges when two graphene layers are stacked with a relative twist angle.[21] The interference between their lattices generates a long-wavelength moiré superlattice, fundamentally modifying the electronic band structure. At a particular "magic" angle, $\theta \approx 1.1^\circ$, interlayer hybridization strongly renormalizes the Dirac velocity, leading to the formation of nearly flat electronic bands.

$$vF(\theta) \to 0 \tag{7.19}$$

In this flat-band regime, the kinetic energy scale becomes strongly suppressed, and electron–electron interactions dominate the physics.[21,22] Quantitatively, the hierarchy $U \gg W$, is established, where U denotes the Coulomb interaction strength and W the bandwidth. Under these conditions, twisted bilayer graphene realizes an effective Hubbard-type model, providing a highly tunable platform for strongly correlated quantum matter.

$$H = -t\sum_{(i,j)} c_i^{\dagger} c_j + U\sum_{(i)} n_{i\uparrow} n_{i\downarrow} \tag{7.20}$$

Experimentally, such systems exhibit correlated insulating states at integer moiré fillings, unconventional superconductivity upon doping, and in certain regimes, topological bands characterized by non-zero Chern numbers. The coexistence of correlation and topology makes moiré systems particularly significant for quantum information science. From a broader perspective, twisted bilayer graphene functions as a solid-state quantum simulator: its Hamiltonian parameters—carrier density, displacement field, twist angle—can be continuously tuned in situ. This level of control enables experimental exploration of many-body phases that are computationally intractable using classical methods. Moreover, topological flat bands in moiré systems provide a promising route toward fractionalized excitations and potentially non-Abelian quasiparticles. If such excitations can be stabilized and controlled, they would offer a pathway toward topological quantum encoding, where logical information is stored nonlocally and thus intrinsically protected from local decoherence.

7.3.2 Diamond and Quantum Computing

Diamond is unique for its room-temperature stability. It acts as a bridge between quantum computing and quantum sensing. Diamond has emerged as a versatile solid-state platform for quantum technologies, distinguished by the presence of optically active color centers embedded in a wide-bandgap host crystal. Unlike superconducting circuits that rely on collective macroscopic degrees of freedom, or semiconductor quantum dots based on confined charge carriers, diamond-based quantum systems are built upon atomically localized defect states whose spin degrees of freedom can be coherently manipulated and optically read out.

Among these defects, the nitrogen-vacancy (NV) center[23–25] has become the most extensively studied due to its remarkable coherence properties, optical accessibility, and capability of operating at room temperature. These characteristics enable diamond to bridge quantum computing, quantum sensing, and quantum networking within a single materials platform. In addition to demonstrating long spin coherence times, diamond NV centers can function as quantum memories, nanoscale sensors, and spin–photon interfaces for distributed quantum networks. At the same time, significant efforts have been devoted to improving fabrication techniques and addressing scalability challenges.

In the following sections, we discuss the physics of color centers in diamond, their role in room-temperature quantum technologies, their potential for quantum memory and networking applications, and current progress toward scalable fabrication.

7.3.2.1 Color Centers in Diamond

At its core, a color center in a diamond can be viewed as a localized quantum system embedded within a crystalline environment. From the perspective of quantum information theory, it comprises a finite-dimensional Hilbert space coupled to structured bosonic and spin baths. The utility of diamond color centers arises not only from their optical visibility but also from their ability to coherently address, manipulate, and measure internal degrees of freedom with high fidelity.

Localized Defect States as Artificial Atoms–Diamond, a wide-bandgap semiconductor with an energy gap $Eg = 5.47\,\text{eV}$, exhibits extended Bloch bands free from defect states in the absence of imperfections. The introduction of a point defect breaks translational symmetry, generating localized electronic states within the bandgap. These states are spatially confined around the defect site and spectrally isolated from the conduction and valence bands.

To first approximation, a color center can be modeled as an "artificial atom" with its electronic structure determined by the symmetry of the defect and the surrounding crystal field. However, unlike real atoms, these artificial atoms are embedded in a solid-state environment, leading to controlled but unavoidable coupling to phonons, nuclear spins, and electromagnetic modes. The total Hamiltonian for this system can be expressed as follows:

$$H = H_{\text{defect}} + H_{\text{bath}} + H_{\text{int}} \tag{7.21}$$

where H_{defect} describes the internal electronic and spin structure of the defect, Hbath represents environmental degrees of freedom (phonons and surrounding spins), and Hint captures system–environment coupling.

The central question for quantum information applications is whether H_{defect} admits a well-isolated, controllable two-level (or few-level) subsystem whose coherent dynamics dominate over dissipative processes induced by H_{int}.

The most extensively studied example is the negatively charged NV^{-}center. It consists of a substitutional nitrogen atom adjacent to a vacancy, forming a defect with C_{3v} symmetry along a $\langle 111 \rangle$ crystallographic axis. Group-theoretic analysis of the electronic orbitals under this symmetry reveals that the ground state is a spin triplet.

Projecting onto the electronic ground manifold yields an effective spin-1 Hamiltonian of the form

$$H_{\text{eff}} = DS_z^2 + E\left(S_x^2 - S_y^2\right) + \gamma_e \boldsymbol{B} \cdot \boldsymbol{S}$$

where D denotes the zero-field splitting arising from dipolar spin–spin interaction, E encodes strain-induced symmetry breaking, and γ_e is the electron gyromagnetic ratio. The zero-field splitting term DS_z^2 lifts the degeneracy between $m_s = 0$ and

$m_s = \pm 1$ even in the absence of an external magnetic field. From a quantum information viewpoint, this intrinsic level splitting provides spectral addressability and protects the logical subspace from low-frequency magnetic noise. By selecting two spin projections (e.g., $m_s = 0$ and $m_s = -1$), one obtains an effective qubit Hilbert space $\mathcal{H}_2 \subset \mathcal{H}_3$. The remaining level may be treated as leakage space, whose effects can be analyzed within standard quantum control theory.

A defining property of many diamond color centers is the existence of spin-dependent optical transitions. The electronically excited states couple to the ground manifold via dipole-allowed transitions, while intersystem crossing through metastable singlet states induces non-unitary spin-selective decay. Operationally, optical excitation implements a non-unitary quantum map that preferentially pumps population into a particular spin sublevel. In density matrix language, optical initialization corresponds to a dissipative map

$$\rho \rightarrow \varepsilon_{\text{opt}}(\rho)$$

whose fixed point approximates a pure spin state. Measurement is realized through spin-dependent fluorescence intensity, effectively implementing a weak measurement in the computational basis. Thus, color centers provide a rare example of a solid-state system in which initialization, coherent control, and measurement are all accessible within a single physical platform.

The electronic spin couples via hyperfine interaction to nearby nuclear spins:

$$H_{hf} = \mathbf{S} \cdot \mathbf{A} \cdot \mathbf{I}$$

These nuclear spins possess significantly longer coherence times due to their weaker magnetic moments. Consequently, the combined electron–nuclear system forms a hybrid quantum register. From a theoretical standpoint, this structure naturally realizes a small quantum processor in which the electronic spin acts as a control and readout qubit, while nuclear spins serve as long-lived quantum memory. Conditional gate operations arise from the dependence of nuclear precession frequencies on the electronic spin state, enabling entangling dynamics within a well-defined few-qubit Hilbert space.

Despite their robustness, color centers remain open quantum systems. The dominant environmental couplings include magnetic dipolar interaction with surrounding ^{13}C nuclear spins, phonon-induced modulation of orbital wavefunctions, and charge-state fluctuations affecting the electronic configuration. These interactions generate effective dephasing and amplitude-damping channels. In many experimental regimes, the magnetic noise can be approximated as quasi-static, leading to Gaussian dephasing dynamics. Dynamical decoupling sequences effectively average out such noise, extending coherence times toward intrinsic limits set by higher-frequency fluctuations. The relatively weak coupling to thermal phonons, even at room temperature, is a direct consequence of diamond's rigid lattice and large bandgap. In the language of open quantum systems, the spectral density of environmental noise is sufficiently suppressed at relevant transition frequencies to permit long-lived coherence.

While the NV center is paradigmatic, other defect families, particularly group-IV vacancy centers (SiV, GeV, and SnV), possess inversion symmetry that reduces first-order sensitivity to electric field noise. Their optical transitions exhibit narrow line-widths and high emission into the zero-phonon line, making them attractive for quantum networking applications. The diversity of defect symmetries provides a broader theoretical landscape in which one may engineer Hamiltonians with tailored noise susceptibilities. In this sense, diamond offers not merely a single-qubit realization, but a materials platform for designing artificial atoms with specified symmetry and environmental coupling properties.

From a quantum information standpoint, color centers in diamond exemplify a central theme: coherence is not determined solely by temperature, but by the interplay between system energy scales, environmental spectral density, and symmetry protection. The defect constitutes a controllable few-level quantum system whose isolation from extended electronic states is guaranteed by the wide bandgap, while its interaction with the lattice remains sufficiently weak to preserve coherence over operational timescales. Thus, color centers should be understood not simply as defects that emit light but as engineered open quantum systems in which both coherent and dissipative dynamics can be harnessed. They form the microscopic substrate upon which room-temperature quantum sensing, memory, and distributed quantum information processing are built.

7.3.2.2 Room-Temperature Quantum Sensing and Computing

The possibility of performing coherent quantum operations at room temperature represents a conceptual departure from the traditional paradigm of quantum information processing. In most quantum platforms, thermal fluctuations rapidly destroy phase coherence, forcing operation in cryogenic environments. Diamond color centers, however, demonstrate that quantum coherence can persist in a solid-state system even at ambient conditions.[26,27] From a theoretical standpoint, this phenomenon raises fundamental questions about decoherence mechanisms, quantum parameter estimation limits, and the structure of controllable open quantum systems.

At finite temperature, a spin defect embedded in diamond constitutes an open quantum system coupled to multiple environmental baths: phonons, nuclear spins, and paramagnetic impurities. The reduced density matrix $\rho(t)$ of the electronic spin evolves according to a quantum dynamical map

$$\rho(t) = \varepsilon_t\left[\rho(0)\right]$$

which can often be approximated as a completely positive trace-preserving map generated by a quantum Lindblad master equation.

The remarkable feature of diamond is that the dominant noise channels at room temperature are relatively weak. The large bandgap suppresses thermally activated charge fluctuations, while the stiffness of the lattice limits phonon-induced spin relaxation. Consequently, the longitudinal relaxation time T_l remains orders of magnitude longer than typical gate times, and the transverse relaxation time T_2 can be extended via dynamical decoupling. In quantum information language, the effective noise strength per gate can be kept below thresholds relevant for fault-tolerant architectures, at least for small-scale registers. The room-temperature regime therefore does not imply the absence of decoherence, but rather that the system resides in a parameter regime where coherent control dominates dissipative dynamics over experimentally relevant timescales.

Quantum sensing with color centers can be formulated rigorously within the framework of quantum parameter estimation theory.[28] Consider a Hamiltonian

$$H(\theta) = H_0 + \theta G$$

where θ is an unknown physical parameter (e.g., magnetic field strength) and G is the generator coupling the system to the field. During an interrogation time τ, the quantum state accumulates a phase

$$\rho(\tau) = e^{-iH(\theta)\tau} \rho(0) e^{iH(\theta)\tau}$$

The precision with which θ can be estimated is bounded by the quantum Cramé r–Rao inequality

$$\Delta\theta \geq \frac{1}{\sqrt{\nu F_Q}}$$

where ν is the number of repetitions and F_Q is the quantum Fisher information. For pure states evolving unitarily,

$$F_Q = 4\tau^2 (\Delta G)^2$$

In the presence of decoherence, the dynamical map becomes non-unitary, and the quantum Fisher information is reduced. The trade-off between phase accumulation and decoherence determines an optimal interrogation time. For diamond spin defects at room temperature, the relatively long T_2 allows substantial phase accumulation before coherence is lost, enabling sensitivities that approach the fundamental quantum limits. Importantly, ensembles of non-interacting defects provide a standard quantum limit scaling $\Delta\theta \sim 1/\sqrt{N}$, while entangled states in principle enable Heisenberg scaling $\sim 1/N$. Whether room-temperature entanglement can be stabilized sufficiently to surpass the standard quantum limit remains an open theoretical and experimental question.

From the perspective of quantum computing, the electronic spin of a color center forms a controllable two-level subsystem embedded in a larger Hilbert space. Microwave driving implements rotations generated by Pauli operators, providing universal single-qubit control:

$$H_{\text{drive}}(t) = \Omega t \cos(\omega t + \phi) Sx$$

Coupling to nearby nuclear spins yields a natural multi-qubit register described by a hyperfine Hamiltonian

$$H_{hf} = \mathbf{S} \cdot \mathbf{A} \cdot \mathbf{I}$$

Conditional dynamics arising from this interaction allow the construction of entangling gates between electron and nuclear qubits. Since single-qubit rotations combined with any entangling two-qubit gate form a universal gate set, small-scale quantum processors are achievable even at room temperature.

However, scalability requires inter-node entanglement. In a distributed architecture, spin–photon entanglement is generated locally, and photonic interference combined with measurement induces remote spin–spin entanglement. Theoretically, this corresponds to entanglement swapping mediated by flying qubits. The success probability and fidelity are highly sensitive to photon indistinguishability, which is partially degraded by phonon interactions at room temperature. Understanding this limitation requires modeling the defect as a spin-boson system with temperature-dependent spectral density.

Room-temperature operation introduces both Markovian and non-Markovian noise channels. Magnetic noise from surrounding nuclear spins often produces slowly varying (quasi-static) fluctuations, which can be mitigated through dynamical decoupling sequences. Phonon-induced relaxation contributes to amplitude-damping channels. Charge-state fluctuations introduce stochastic jumps between Hilbert space sectors.

In the language of quantum error correction, these processes correspond primarily to phase-flip and amplitude-damping errors. The feasibility of fault tolerance depends on achieving logical error rates below threshold values determined by the chosen code. While large-scale fault-tolerant quantum computing in diamond remains a long-term goal, theoretical analyses indicate that small logical qubits encoded in electron-nuclear registers are within reach.

In summary, the broader theoretical importance of room-temperature diamond systems lies in their demonstration that quantum coherence is not synonymous with cryogenic isolation. Instead, coherence can survive whenever the system–environment coupling is sufficiently weak relative to control energy scales. This shifts the emphasis from temperature alone to spectral properties of the environment and symmetry-protected defect structures. In this sense, diamond color centers provide a concrete realization of a controllable open quantum system operating far from the zero-temperature limit. They serve as a testbed for theories of decoherence, quantum metrology in the presence of noise, and distributed quantum information processing. Room-temperature quantum sensing and computing in diamond, therefore, represent not merely an engineering convenience but a conceptual expansion of the domain in which quantum information theory can be physically implemented.

7.3.2.3 Quantum Memory and Networking

Quantum memory[29] and quantum networking[30] represent two logically distinct, yet physically intertwined, extensions of single-defect quantum control. If a color center provides a controllable few-level system, then quantum memory concerns the faithful preservation of quantum states over time, while quantum networking concerns the faithful distribution of quantum states across space. Both tasks are fundamentally constrained by decoherence, noise channels, and the structure of light–matter interaction.

A quantum memory can be formally described as a quantum channel Mt acting over a storage time t,

$$\rho(0) \mapsto \rho(t) = \mathcal{M}_t\left[\rho(0)\right]$$

The quality of a memory is characterized by its ability to preserve coherence and entanglement. In information–theoretic terms, one may quantify this ability via the entanglement fidelity, the diamond norm distance to the identity channel, or the quantum capacity of $\mathcal{M}_t$.

For an electronic spin in a color center, the dominant noise mechanisms at room temperature typically reduce to dephasing and amplitude-damping channels. In the simplest approximation, pure dephasing leads to

$$\rho_{01}(t) = \rho_{01}(0)e^{-t/T_2}$$

while populations remain unchanged. The memory lifetime is therefore directly tied to the coherence time T_2.

However, diamond color centers possess a crucial hierarchical structure: nearby nuclear spins exhibit coherence times that exceed those of the electron spin by several orders of magnitude. Through hyperfine interaction,

$$H_{hf} = \mathbf{S} \cdot \mathbf{A} \cdot \mathbf{I}$$

Quantum information can be coherently transferred from the electronic spin to a nuclear spin. Operationally, this implements a SWAP-like unitary within the joint Hilbert space,

$$U_{\text{swap}} : H_e \otimes H_n \to H_e \otimes H_n$$

The nuclear spin then functions as a long-lived quantum memory, while the electron spin remains available for communication and readout. The resulting architecture is naturally interpreted as a small quantum register in which different subsystems have different noise profiles. From the perspective of quantum error correction, this heterogeneity is advantageous: fast but noisy qubits (electrons) can interface with slow but stable qubits (nuclei).

The performance of a quantum memory is ultimately limited by environmental couplings. Magnetic dipolar interactions with surrounding nuclear spins generate slowly varying phase noise, often modeled as a non-Markovian bath. Phonon-induced relaxation introduces amplitude damping. Dynamical decoupling sequences effectively engineer the system–environment interaction by averaging out low-frequency noise components. In the filter-function formalism, the effective decoherence rate becomes

$$\Gamma_{\text{eff}} = \int_0^{\infty} S(\omega)\left|F(\omega)\right|^2 d\omega$$

where $S(\omega)$ is the environmental spectral density and $F(\omega)$ is the control-dependent filter function. By shaping $F(\omega)$ appropriately, one suppresses the dominant noise frequencies, thereby extending storage time. Thus, quantum memory in diamond is not a passive property but an actively engineered feature of an open quantum system.

While quantum memory concerns temporal coherence, quantum networking concerns spatial coherence. In diamond color centers, networking relies on entanglement between stationary spin qubits and flying photonic qubits. Consider a defect initially prepared in a superposition state

$$|\psi\rangle = \alpha|0\rangle + \beta|1\rangle$$

Spin-dependent optical transitions generate entanglement between the spin state and the emitted photon polarization or time bin,

$$|\Psi\rangle = \alpha|0\rangle|1_{\gamma},0\rangle + \beta|1\rangle|1_{\gamma},1\rangle$$

If two spatially separated defects emit photons that interfere on a beamsplitter, conditional detection of specific photonic outcomes projects the remote spins into an entangled state. This procedure realizes measurement-induced entanglement swapping, forming the basis of a quantum repeater architecture. From a theoretical standpoint, the process can be described as local spin–photon entanglement generation (unitary light–matter coupling), photonic interference (linear optical unitary), and conditional measurement (non-unitary projection).

Heralded entanglement between remote spins. The success probability is intrinsically probabilistic and depends on photon indistinguishability and collection efficiency. Decoherence during photon emission can be modeled using a spin-boson Hamiltonian with temperature-dependent spectral density, which broadens optical transitions and reduces entanglement fidelity.

Scalable quantum networking requires overcoming photon loss and decoherence over long distances. Quantum repeaters combine entanglement generation with quantum memory. Stored entangled states are extended via entanglement swapping operations across intermediate nodes. The key resource enabling such repeaters in diamond is the coexistence of an optical interface (electronic spin), a long-lived storage qubit (nuclear spin), and local entangling gates (hyperfine coupling).

In this hybrid architecture, entanglement is first established between electron spins of neighboring nodes. It is then transferred to nuclear memories for long-term storage while subsequent entanglement links are generated. Theoretical performance is evaluated in terms of entanglement distribution rate, fidelity, and memory lifetime relative to photon transmission delay.

The ultimate limits of quantum networking in diamond are governed by quantum channel capacities. Photon loss in optical fibers constitutes a bosonic attenuation channel with transmissivity η. Without repeaters, the secret-key and quantum capacities decay exponentially with distance. The presence of long-lived quantum memories at network nodes allows one to surpass direct transmission limits by dividing the channel into shorter segments. However, the practical advantage depends critically on memory coherence time exceeding classical communication latency between nodes. Thus, the interplay between storage time, entanglement generation rate, and photon loss determines whether a diamond-based network operates below or above the repeater threshold regimes.

In summary, quantum memory and networking in diamond color centers illustrate a unifying principle of quantum information theory: information is preserved or distributed only insofar as coherence can be protected against environmental coupling. Diamond's material properties provide an unusual regime in which optical addressability, long-lived nuclear memory, and room-temperature robustness coexist within a single physical platform. In this sense, color centers are not merely qubits but nodes in a distributed open quantum system. They enable the experimental realization of fundamental concepts such as entanglement swapping, heralded state preparation, hybrid quantum registers, and quantum channel engineering. As such, diamond-based quantum memory and networking architectures occupy a central position in the broader program of scalable quantum information processing.

7.3.2.4 Fabrication and Scalability

The question of fabrication in diamond-based quantum technologies should not be understood merely as a materials engineering challenge.[31,32] From the perspective of quantum information theory, fabrication concerns the reproducible realization of controllable quantum systems whose Hamiltonians, noise channels, and interconnections satisfy the structural requirements of scalable architectures. Scalability, in turn, is not defined by the number of qubits alone, but by the ability to maintain control fidelity and coherence as system size grows.

A single color center may be modeled as a localized few-level system with an effective Hamiltonian

$$H_{\text{eff}} = H_{\text{internal}} + H_{\text{control}}\left(t\right) + H_{\text{noise}}$$

For large-scale quantum information processing, it is necessary that many such systems be fabricated with nearly identical internal Hamiltonians. Variations in zero-field splitting, strain-induced terms, or hyperfine couplings translate into inhomogeneous qubit frequencies and gate errors. Inhomogeneity can be treated theoretically as static disorder in the system Hamiltonian. While moderate disorder may be compensated through calibration and optimal control, large disorder increases control overhead and may render uniform gate sequences ineffective. Therefore, fabrication precision directly affects the complexity of control protocols required to implement a universal gate set. From a theoretical standpoint, scalability demands that fabrication-induced parameter fluctuations remain within a regime where they can be absorbed into local unitary corrections rather than fundamentally altering the system's spectral structure.

Moreover, scalable quantum computing requires controlled interactions between qubits. In diamond platforms, interactions arise either from magnetic dipolar coupling or photonic mediation. Deterministic Placement of color centers, therefore, determines the interaction graph of the quantum processor. If qubits are randomly distributed, the resulting coupling graph is stochastic and difficult to exploit algorithmically. In contrast, Deterministic Placement allows one to design a desired connectivity graph G(V, E), where vertices correspond to qubits and edges correspond to controllable interactions. From a quantum circuit perspective, connectivity constraints determine the compilation overhead required to implement arbitrary two-qubit gates. Limited connectivity increases circuit depth through SWAP operations, thereby amplifying exposure to decoherence. Thus, fabrication precision directly influences effective logical error rates.

For distributed architectures, scalability depends critically on the efficiency of spin–photon interfaces. Photon collection efficiency η enters exponentially into entanglement generation rates across a network. Low η reduces the probability of successful heralded entanglement and increases idle time for quantum memories. Nanophotonic structures—such as cavities and waveguides—modify the local photonic density of states, enhancing emission into the zero-phonon line via the Purcell effect. Theoretically, this corresponds to engineering the system–bath coupling so that the desired radiative channels dominate over the undesired dissipative ones. In this sense, fabrication is equivalent to shaping the environmental spectral density. Scalability demands that this shaping be reproducible across many nodes, ensuring uniform light–matter coupling strengths.

Concerning realistic applications, any scalable quantum architecture must confront the threshold theorem of quantum error correction.[4–7] Let p denote the physical error probability per gate. Fault-tolerant computation is possible only if $p < p\text{th}$, where pth depends on the chosen error-correcting code and noise model. In diamond systems, dominant physical error channels include dephasing due to magnetic noise, amplitude damping due to phonon-induced relaxation, photon loss in networking operations, and leakage outside the computational subspace.

Fabrication quality influences each of these channels. Impurity concentration affects magnetic noise; strain and charge instability affect optical linewidths; nanofabrication precision determines photon collection efficiency. Thus, fabrication precision directly enters into the effective noise parameters of the quantum channel. Scalability requires that these parameters lie within the basin of attraction of a fault-tolerant code.

Given the difficulty of achieving strong, deterministic long-range interactions in solid-state systems, a modular architecture offers a natural scalability pathway. Each module consists of a small, high-fidelity quantum register (e.g., an electron spin coupled to several nuclear spins). Modules are connected probabilistically through photonic links. In such an architecture, scalability reduces to three quantitative conditions: (i) Memory coherence time must exceed inter-module communication latency, (ii) Entanglement generation rate must exceed decoherence rate, and (iii) Local gate fidelity must remain above threshold under realistic disorder. Fabrication advances that improve any one of these parameters expand the feasible size of the network.

As device dimensions shrink, surface effects become increasingly important. Surface states introduce electric field fluctuations that broaden optical transitions and destabilize charge states. Theoretically, these effects modify the system–environment interaction Hamiltonian and introduce additional dephasing channels. Surface termination and passivation can be interpreted as techniques for engineering boundary conditions in the open-system model. Proper surface control reduces low-frequency noise components and enhances spectral stability, directly improving entanglement fidelity in networking protocols.

Ultimately, scalability in diamond quantum technologies is a resource problem. The relevant resources include: (i) Coherence time relative to gate time, (ii) Optical interface efficiency, (iii) Deterministic connectivity, and (iv) Fabrication reproducibility. Each resource can be mapped onto information–theoretic quantities: channel fidelity, entanglement rate, and logical error probability. Fabrication is therefore not merely a technological step preceding quantum information processing; it is an intrinsic component of the physical realization of abstract quantum channels and operations.

The fabrication and scalability problem in diamond systems highlights a broader principle: a quantum information architecture is defined as much by its noise structure as by its ideal Hamiltonian. Engineering a scalable platform means engineering both coherent dynamics and dissipative processes across many identical units. Diamond color centers offer a uniquely promising route because their microscopic structure already satisfies several essential criteria: spectral isolation due to the wide bandgap, room-temperature coherence, and optical addressability. Continued advances in deterministic defect placement, nanophotonic integration, and surface control move the platform closer to a regime where the assumptions of fault-tolerant quantum information theory can be physically realized. In this framework, fabrication is best understood as the process of constructing identical, controllable open quantum systems whose collective behavior supports large-scale quantum information processing.

7.3.3 Low-Dimensional Topological Insulators and Quantum Computing

Topological computing aims for "Hardware-level" error protection via Majorana fermions and topological effects,[33–35] which are immune to local perturbations. As such, topological computing introduces a fundamentally different approach to quantum information processing. Rather than relying solely on material purity or dynamical decoupling techniques to preserve coherence, topological systems aim to encode quantum information in global, nonlocal properties of the many-body wavefunction that are intrinsically protected against local perturbations.

In low-dimensional systems, topology manifests through protected boundary states whose existence is guaranteed by bulk topological invariants. Notably, 2D topological insulators host helical edge modes immune to backscattering under time-reversal symmetry, while one-dimensional (1D) topological superconductors support zero-energy

Majorana bound states at their boundaries. These exotic quasiparticles provide a platform for fault-tolerant quantum computation based on non-Abelian statistics and topological braiding operations. From a theoretical perspective, topological quantum computing shifts the focus from microscopic Hamiltonian engineering to the manipulation of degenerate ground-state manifolds characterized by topological order. The key question is not merely how to suppress decoherence, but how to encode qubits in degrees of freedom that are inherently insensitive to local noise.

In the following sections, we introduce the concept of topological protection, examine 2D quantum spin Hall (QSH) systems and topological superconducting nanowires, discuss the implementation of topological quantum gates, and review recent experimental progress toward realizing scalable topological quantum processors.

7.3.3.1 Topological Protection

Topological phases of matter are characterized not by local order parameters, but by global topological invariants defined over momentum space or many-body Hilbert space. Unlike conventional phases distinguished by spontaneous symmetry breaking, topological phases remain stable under continuous deformations of the Hamiltonian so long as the protecting symmetry and energy gap are preserved.

Consider a gapped Hamiltonian $H(\lambda)$ depending smoothly on parameters λ. Two Hamiltonians belong to the same topological phase if they can be connected without closing the bulk energy gap. The robustness of topological phases therefore arises from the impossibility of removing gapless boundary modes without inducing a bulk phase transition. In non-interacting fermionic systems, topological classification can be formulated in terms of symmetry class and spatial dimension, leading to the so-called tenfold way. The relevant invariants include Chern numbers, Z_2 indices, and winding numbers. For example, in 2D systems with broken time-reversal symmetry, the Chern number

$$C = \frac{1}{2\pi}\int_{BZ} \Omega(k)\, d^2k$$

quantifies the topological structure of occupied bands via the Berry curvature $\Omega(k)$.

Topological protection emerges because local perturbations cannot change these global invariants. In quantum information terms, this implies that certain degrees of freedom, such as boundary modes, are insensitive to local noise, provided the system remains within the same topological phase. The protection mechanism is therefore fundamentally geometric rather than energetic. When extended to systems supporting non-Abelian quasiparticles, topological protection acquires direct computational meaning: quantum information encoded nonlocally in topological degrees of freedom becomes immune to local decoherence processes.

7.3.3.2 2D Topological Insulators (Quantum Spin Hall Insulators)

2D topological insulators, also known as QSH insulators,[36–38] are characterized by an insulating bulk and gapless helical edge states protected by time-reversal symmetry. A minimal model is given by the Bernevig–Hughes–Zhang (BHZ) Hamiltonian,[37]

$$H(k) = \begin{pmatrix} h(k) & 0 \\ 0 & h(k) \end{pmatrix}$$

where

$$h(k) = \left(M - Bk^2\right)\sigma_z + A\left(k_x\sigma_x + k_y\sigma_y\right)$$

The sign of M / B determines whether the system is topologically trivial or non-trivial. The Z_2 invariant distinguishes these phases, reflecting the parity of band inversions.

At the boundary, gapless helical modes emerge. These modes exhibit spin-momentum locking: electrons with opposite spins propagate in opposite directions. Backscattering from non-magnetic disorder is suppressed by time-reversal symmetry, since it would require a spin-flip. From a quantum information perspective, the helical edge channels form robust 1D conducting channels with reduced sensitivity to disorder. However, since the protection relies on symmetry rather than on nonlocal encoding, QSH systems do not, by themselves, provide topological qubits. Their significance lies in forming building blocks for proximity-induced superconductivity and Majorana physics.

7.3.3.3 Topological Superconducting Nanowires

Notably, 1D semiconductor nanowires with strong SOC, when coupled to an s-wave superconductor and subjected to a Zeeman field, can enter a topological superconducting phase supporting Majorana zero modes at their ends.[39–41] A minimal Bogoliubov–de Gennes Hamiltonian is

$$H(k) = \left(\frac{k^2}{2m} - \mu\right)\tau_z + \alpha k\sigma_y\tau_z + V_Z\sigma_x + \Delta\tau_x$$

where σi act on spin space and τi act on particle–hole space. When the condition

$$V_Z^2 > \mu^2 + \Delta^2$$

is satisfied, the system enters a topological phase characterized by a nontrivial winding number. In this phase, zero-energy Majorana modes γ_1 and γ_2 localize at opposite wire ends. Majorana operators satisfy

$$\gamma_i = \gamma_i^\dagger, \{\gamma_i, \gamma_j\} = 2\delta_{IJ}$$

A pair of Majorana modes defines a nonlocal fermionic mode

$$c = 1/2\left(\gamma_1 + i\gamma_2\right)$$

Quantum information encoded in the occupation parity of this fermionic mode is inherently nonlocal. Local perturbations acting at one end cannot change global fermion parity, providing topological protection against local decoherence.

7.3.3.4 Topological Quantum Gates

The most striking implication of Majorana zero modes is their non-Abelian exchange statistics.[42–44] When two Majorana modes are adiabatically exchanged (braided), the system undergoes a unitary transformation

$$\gamma_i \to \gamma_j, \gamma_i \to -\gamma_i$$

In the degenerate ground-state manifold, braiding implements a non-commuting unitary transformation:

$$U_{ij} = \exp\left(\frac{\pi}{4}\gamma_i\gamma_j\right)$$

These operations depend only on the topology of the exchange path, not on microscopic details, giving rise to fault-tolerant quantum gates.

However, braiding of Majorana modes generates only a subset of unitary operations (the Clifford group). Universal quantum computation requires supplementation with non-topological gates, such as the $\pi/8$ phase gate, typically implemented through magic-state distillation. Thus, topological quantum computing combines inherently fault-tolerant operations with conventional error-corrected gates to achieve universality.

7.3.3.5 Experimental Frontiers

Experimental efforts focus on detecting unambiguous signatures of Majorana zero modes and achieving controllable braiding operations.[45–48] Zero-bias conductance peaks observed in proximitized nanowires are consistent with Majorana physics, although alternative explanations remain under investigation. Beyond semiconductor nanowires, platforms under active investigation include 2D topological insulator edges proximitized by superconductors, magnetic atom chains on superconductors, and hybrid superconductor–topological insulator heterostructures.

The central theoretical challenge remains bridging the gap between proof-of-principle Majorana detection and scalable, fault-tolerant topological quantum computation. In the experimental pursuit of Majorana zero modes, one also faces several key challenges: (i) How to distinguish true topological zero modes from trivial Andreev bound states? (ii) How to demonstrate non-Abelian braiding statistics? (iii) How to increase the topological gap to enhance robustness? (iv) How to integrate scalable wire networks?

7.3.4 Low-Dimensional Superconducting Materials and Quantum Computing

Superconducting computing utilizing superconducting qubits is currently the most “scalable” platform. Superconducting systems provide a distinct route toward quantum computation by exploiting macroscopic quantum coherence in condensed matter systems. Unlike localized defect spins in diamond or confined charge carriers in semiconductor quantum dots, superconducting qubits are constructed from collective phase degrees of freedom of Cooper pairs. These collective variables behave as nonlinear quantum oscillators that can be engineered, coupled, and controlled using lithographically fabricated circuits.

Low-dimensional superconducting materials, particularly 2D van der Waals superconductors, have expanded the design space of superconducting quantum devices. Reduced dimensionality modifies pairing symmetry, enhances tunability via electrostatic gating, and enables the integration of superconductivity with other quantum materials. This material's flexibility complements the circuit-based architecture that underlies superconducting quantum processors.

From a theoretical standpoint, superconducting quantum circuits realize engineered open quantum systems in which the Hamiltonian, nonlinearity, and coupling topology can be designed at the fabrication stage. The central challenges are therefore twofold: constructing qubit architectures with strong anharmonicity and controllable interactions, and mitigating decoherence arising from material defects, dielectric loss, and environmental fluctuations.

In the following sections, we discuss 2D van der Waals superconductors as emerging material platforms, examine the principles of superconducting qubit architectures, analyze dominant noise mechanisms and mitigation strategies, and explore hybrid systems that combine superconductivity with other quantum degrees of freedom.

7.3.4.1 2D van der Waals Superconductors

Low-dimensional superconductors represent a regime in which reduced dimensionality enhances quantum fluctuations while preserving macroscopic phase coherence. In 2D van der Waals superconductors,[21,49–51] superconductivity emerges in atomically thin layers that are weakly coupled by van der Waals forces. This structural flexibility allows heterostructure engineering at the level of single atomic layers.

At the microscopic level, superconductivity is described by the Bardeen–Cooper–Schrieffer Hamiltonian,

$$H_{\mathrm{BCS}} = \sum_{k,\sigma} \varepsilon_k c_{k\sigma}^{\dagger} c_{k\sigma} - \sum_{k,k'} V_{kk'} C_{k\uparrow}^{\dagger} C_{-k\uparrow}^{\dagger} C_{-k'\downarrow} C_{k'\uparrow}$$

The superconducting order parameter

$$\Delta = V \sum_{k} \left\langle c_{-k\downarrow} c_{k\uparrow} \right\rangle$$

introduces a complex phase ϕ, whose fluctuations dominate low-energy dynamics.

In two dimensions, long-range phase order at finite temperature is governed by the Berezinskii–Kosterlitz–Thouless mechanism rather than by conventional symmetry breaking. The superconducting transition corresponds to vortex–antivortex unbinding. This highlights an essential feature of low-dimensional superconductors: amplitude fluctuations are typically suppressed by the pairing gap, while phase fluctuations determine coherence properties. For quantum information applications, the relevant low-energy degree of freedom is the macroscopic superconducting phase. The conjugate variable to phase is the Cooper pair number, satisfying

$$\left[\hat{\phi},\hat{n}\right]=i$$

This canonical pair forms the foundation of superconducting qubit Hamiltonians.

7.3.4.2 Superconducting Qubit Architectures

Superconducting qubits arise from nonlinear circuit elements constructed from Josephson junctions.[52–55] A Josephson junction provides a non-dissipative nonlinear inductance described by the Hamiltonian

$$H=4E_C\hat{n}^2-E_J\cos\hat{\phi}$$

where $E_C=e^2/2C$ is the charging energy, and EJ is the Josephson energy. The cosine term introduces anharmonicity, enabling the lowest two energy levels to define a qubit. Depending on the ratio E_J/E_C, different qubit modalities emerge: (i) Charge qubits satisfying $E_C \gg E_J$, (ii) Flux qubits where magnetic flux serves as the primary variable, (iii) Phase qubits, and (iv) Transmon qubit with $E_J/E_C \gg 1$ where charge noise is largely suppressed. In the working regime of the Transmon qubit, the Hamiltonian approximates a weakly anharmonic oscillator. The qubit frequency is

$$\omega_{01}\approx\sqrt{8E_JE_C}-E_C$$

Superconducting qubits are coupled via capacitive or inductive interactions, leading to effective two-qubit Hamiltonians such as

$$H_{\text{int}}=g\sigma_x^{(1)}\sigma_x^{(2)}$$

Because all interactions are engineered at the circuit level, superconducting systems realize a programmable quantum many-body Hamiltonian. From a quantum information standpoint, they provide fast gate times (nanoseconds) and strong controllability, although at the cost of requiring cryogenic operation. The Hilbert space of a superconducting processor therefore emerges from collective phase degrees of freedom in a macroscopic quantum state—a striking example of emergent quantum information carriers.

7.3.4.3 Noise Mitigation

Despite strong controllability, superconducting qubits are inherently open quantum systems. Dominant noise channels include,[56–58] charge noise from fluctuating offset charges, flux noise from magnetic impurities, dielectric loss due to TLSs, and quasi-particle poisoning.

These mechanisms contribute to dephasing and energy relaxation, characterized by times T_2 and T_1. To mitigate these noises, one usually implements the following strategies.

Hamiltonian design: Increasing E_J / E_C reduces sensitivity to charge noise.
Materials engineering: Improving substrate purity reduces TLS density.
Dynamical decoupling: Control pulses suppress low-frequency noise.
Quantum error correction: Logical encoding reduces effective error rates below threshold.

In modern architectures, logical qubits are encoded in surface codes requiring 2D nearest-neighbor connectivity. The error threshold for surface codes lies near 10–2 under favorable noise models, motivating continual reduction of physical error probabilities. Thus, noise mitigation is both a materials challenge and an information–theoretic optimization problem.

7.3.4.4 Hybrid Systems

Hybrid quantum systems combine superconducting circuits with other quantum platforms to leverage complementary strengths.[59,60] Examples include coupling superconducting qubits to (i) Spin ensembles, (ii) Mechanical resonators, (iii) Photonic cavities, or (iv) Topological modes.

In circuit quantum electrodynamics, a superconducting qubit couples to a microwave resonator via a Jaynes–Cummings Hamiltonian,

$$H = \omega_r a^\dagger a + \frac{\omega q}{2}\sigma_z + g\left(a\sigma^+ + a^\dagger\sigma^-\right) \tag{7.22}$$

This architecture enables long-distance coupling between qubits and high-fidelity readout.

Hybridization with spin systems (e.g., NV centers) aims to combine fast superconducting logic with long-lived spin memory. The effective interaction can be modeled as a mediated coupling through a resonator mode. More speculative hybrid proposals involve coupling superconductors to topological materials, seeking to merge fast circuit-based control with topologically protected degrees of freedom. From a theoretical viewpoint, hybrid systems expand the accessible Hilbert space and diversify noise spectra. They represent an attempt to engineer composite open quantum systems in which the subsystems compensate for one another's weaknesses.

7.4 CHALLENGES AND FUTURE

As we transition from the NISQ era toward Fault-Tolerant Quantum Computing (FTQC), several critical bottlenecks must be addressed.

7.4.1 The Scaling Wall and Interconnectivity

Current systems struggle to move beyond the few-hundred-qubit range. For superconducting circuits, the heat load from coaxial cables in dilution refrigerators creates a "wiring bottleneck". Future architectures must transition to Cryogenic CMOS controllers or Photonic Interconnects to manage thousands of qubits. In diamond-based systems, the challenge is the Deterministic Placement of NV centers with nanometer precision to allow for consistent dipolar coupling between nodes.

7.4.2 Material Inhomogeneity and Decoherence

No material is perfect. In GQDs, edge roughness and charge impurities cause "potential fluctuations" that lead to qubit variability. In superconducting qubits, TLS at the metal-oxide interfaces remain the primary source of dielectric loss. The future lies in Materials Genomics, using AI and high-throughput simulation to discover substrates with ultra-low loss tangents and high isotopic purity. (iii) The Quest for the Majorana Qubit Topological quantum computing remains the "Holy Grail." While the theory of Majorana Bound States (MBS) is robust, experimental signatures are often confounded by trivial Andreev Bound States. The future requires cleaner heterostructures (e.g., Al/InAs nanowires) and the first unambiguous demonstration of Non-Abelian Braiding, which would prove that a qubit can be protected by topology rather than just active error correction.

7.4.3 Hybrid Quantum Systems

It is unlikely that one single material will "win" the race. The future will likely see Hybrid Architectures including superconducting qubits for fast processing, diamond NV centers or nuclear spins for long-term quantum memory, and photonic circuits for long-distance networking. Integrating these disparate materials on a single chip, i.e., quantum system-on-a-chip, is the next great engineering challenge.

Looking forward, the ultimate goal is a system where the logical error rate is suppressed exponentially by increasing the number of physical qubits. Whether this is achieved through the brute-force scaling of Transmons or the elegant protection of topological phases, the journey will fundamentally redefine our understanding of matter. Quantum computing is not just the future of calculation; it is the ultimate test of our ability to command the laws of nature.

REFERENCES

[1] M. A. Nielsen and I. L. Chuang, *Quantum Computation and Quantum Information* (Cambridge University Press, 2000).

[2] W. K. Wootters and W. H. Zurek, A single quantum cannot be cloned. *Nature* 299, 802 (1982).

[3] H.-P. Breuer and F. Petruccione, *The Theory of Open Quantum Systems* (Oxford University Press, 2002).

[4] P. W. Shor, Fault-tolerant quantum computation. In *Proceedings of 37th Conference on Foundations of Computer Science*, pp. 56–65 (1996).

[5] E. Knill, R. Laflamme, and W. H. Zurek, Resilient quantum computation. *Science* 279, 342 (1998).
[6] A. Y. Kitaev, Fault-tolerant quantum computation by anyons. *Annals of Physics* 303, 2 (2003).
[7] D. Aharonov and M. Ben-Or, Fault-tolerant quantum computation with constant error rate. *SIAM Journal on Computing* 38, 1207 (2008).
[8] F. Motzoi, J. M. Gambetta, P. Rebentrost, and F. K. Wilhelm, Simple pulses for elimination of leakage in weakly nonlinear qubits. *Physical Review Letters* 103, 110501 (2009).
[9] L. Viola and S. Lloyd, Dynamical suppression of decoherence in two-state quantum systems. *Physical Review A* 58, 2733 (1998).
[10] L. Viola, E. Knill, and S. Lloyd, Dynamical decoupling of open quantum systems. *Physical Review Letters* 82, 2417 (1999).
[11] S. Machnes et al., Comparing, optimizing, and benchmarking quantum-control algorithms in a unifying programming framework. *Physical Review A* 84, 022305 (2011).
[12] L. Banszerus et al., Particle–hole symmetry protects spin-valley blockade in graphene quantum dots. *Nature* 618, 51 (2023).
[13] H. Duprez et al., Spin-valley locked excited states spectroscopy in a one-particle bilayer graphene quantum dot. *Nature Communications* 15, 9717 (2024).
[14] N. M. Freitag et al., Electrostatically confined monolayer graphene quantum dots with orbital and valley splittings. *Nano Letters* 16, 5798 (2016).
[15] R. Krishnan et al., Spin-valley locking for in-gap quantum dots in a mos_2 transistor. *Nano Letters* 23, 6171 (2023).
[16] E. Kogan, Graphene for electronics. *Nanomaterials* 12, 4359 (2022).
[17] K. S. Novoselov et al., Two-dimensional gas of massless Dirac fermions in graphene. *Nature* 438, 197 (2005).
[18] M. Park, Y. Kim, and H. Lee, Design of 2d massless Dirac fermion systems and quantum spin hall insulators based on sp–sp^2 carbon sheets. *NPJ Computational Materials* 4, 54 (2018).
[19] S. Das Sarma, S. Adam, E. H. Hwang, and E. Rossi, Electronic transport in two-dimensional graphene. *Reviews of Modern Physics* 83, 407 (2011).
[20] L.-J. Yin et al., Landau quantization of Dirac fermions in graphene and its multilayers. *Frontiers of Physics* 12, 127208 (2017).
[21] Y. Cao et al., Unconventional superconductivity in magic-angle graphene superlattices. *Nature* 556, 43 (2018).
[22] Y. Cao et al., Correlated insulator behaviour at half-filling in magic-angle graphene superlattices. *Nature* 556, 80 (2018).
[23] M. W. Doherty et al., The nitrogen-vacancy colour centre in diamond. *Physics Reports* 528, 1 (2013).
[24] R. Hanson and D. D. Awschalom, Coherent manipulation of single spins in semiconductors. *Nature* 453, 1043 (2008).
[25] F. Jelezko and J. Wrachtrup, Single defect centres in diamond: A review. *Physica Status Solidi (a)* 203, 3207 (2006).
[26] G. Balasubramanian et al., Ultralong spin coherence time in isotopically engineered diamond. *Nature Materials* 8, 383 (2009).
[27] L. Rondin et al., Magnetometry with nitrogen-vacancy defects in diamond. *Reports on Progress in Physics* 77, 056503 (2014).
[28] C. L. Degen, F. Reinhard, and P. Cappellaro, Quantum sensing. *Reviews of Modern Physics* 89, 035002 (2017).
[29] P. C. Maurer et al., Room-temperature quantum bit memory exceeding one second. *Science* 336, 1283 (2012).

[30] H. J. Kimble, The quantum internet. *Nature* 453, 1023 (2008).
[31] E. Neu et al., Photonic nano-structures on (111)-oriented diamond. *Nature Communications* 5, 5869 (2014).
[32] D. A. Broadway et al., Spatial mapping of strain in diamond using widefield NV imaging. *Physical Review Applied* 9, 044042 (2018).
[33] M. Z. Hasan and C. L. Kane, Colloquium: Topological insulators. *Reviews of Modern Physics* 82, 3045 (2010).
[34] X.-L. Qi and S.-C. Zhang, Topological insulators and superconductors. *Reviews of Modern Physics* 83, 1057 (2011).
[35] B. A. Bernevig and T. L. Hughes, *Topological Insulators and Topological Superconductors* (Princeton University Press, 2013).
[36] C. L. Kane and E. J. Mele, Quantum spin Hall effect in graphene. *Physical Review Letters* 95, 226801 (2005).
[37] B. A. Bernevig, T. L. Hughes, and S.-C. Zhang, Quantum spin Hall effect and topological phase transition in HgTe quantum wells. *Science* 314, 1757 (2006).
[38] M. Konig et al., Quantum spin hall insulator state in HgTe quantum wells. *Science* 318, 766 (2007).
[39] A. Y. Kitaev, Unpaired majorana fermions in quantum wires. *Physics-Uspekhi* 44, 131 (2001).
[40] R. M. Lutchyn, J. D. Sau, and S. Das Sarma, Majorana fermions and a topological phase transition in semiconductor-superconductor heterostructures. *Physical Review Letters* 105, 077001 (2010).
[41] Y. Oreg, G. Refael, and F. von Oppen, Helical liquids and Majorana bound states in quantum wires. *Physical Review Letters* 105, 177002 (2010).
[42] C. Nayak, S. H. Simon, A. Stern, M. Freedman, and S. Das Sarma, Non-abelian anyons and topological quantum computation. *Reviews of Modern Physics* 80, 1083 (2008).
[43] J. Alicea, Non-abelian statistics and topological quantum information processing in 1d wire networks. *Reports on Progress in Physics* 75, 076501 (2012).
[44] P. Bonderson, M. H. Freedman, and C. Nayak, Measurement-only topological quantum computation. *Physical Review Letters* 101, 010501 (2008).
[45] V. Mourik et al., Signatures of majorana fermions in hybrid superconductor-semiconductor nanowire devices. *Science* 336, 1003 (2012).
[46] S. Nadj-Perge et al., Observation of Majorana fermions in ferromagnetic atomic chains on a superconductor. *Science* 346, 602 (2014).
[47] S.-Y. Xu et al., Observation of a topological crystalline insulator phase and topological dirac surface states in pb1xsnxse. *Nature Communications* 5, 4566 (2014).
[48] C. Z. Chang et al., Experimental observation of the quantum anomalous hall effect in a magnetic topological insulator. *Science* 340, 167 (2013).
[49] X. Xi et al., Ising pairing in superconducting $Nbse_2$ atomic layers. *Nature Physics* 12, 139 (2016).
[50] Y. Saito, T. Nojima, and Y. Iwasa, Highly crystalline 2d superconductors. *Nature Reviews Materials* 2, 16094 (2016).
[51] M. M. Ugeda et al., Characterization of collective ground states in single-layer nbse2. *Nature Physics* 12, 92 (2016).
[52] M. H. Devoret and R. J. Schoelkopf, Superconducting circuits for quantum information: An outlook. *Science* 339, 1169 (2013).
[53] J. Clarke and F. K. Wilhelm, Superconducting quantum bits. *Nature* 453, 1031 (2008).
[54] J. Koch et al., Charge-insensitive qubit design derived from the cooper pair box. *Physical Review A* 76, 042319 (2007).

[55] M. Kjaergaard et al., Superconducting qubits: Current state of play. *Annual Review of Condensed Matter Physics* 11, 369 (2020).
[56] J. M. Martinis et al., Decoherence in josephson qubits from dielectric loss. *Physical Review Letters* 95, 210503 (2005).
[57] J. Bylander et al., Noise spectroscopy through dynamical decoupling. *Nature Physics* 7, 565 (2011).
[58] F. Yan et al., The flux qubit revisited. *Nature Communications* 7, 12964 (2016).
[59] Z.-L. Xiang et al., Hybrid quantum circuits: Superconducting circuits interacting with other quantum systems. *Reviews of Modern Physics* 85, 623 (2013).
[60] G. Kurizki et al., Quantum technologies with hybrid systems. *PNAS* 112, 3866 (2015).

8 Low-Dimensional Quantum Materials in Novel Electronic and Optoelectronic Devices

Wei Chen, Xiang Qi, and Long Ren

8.1 INTRODUCTION

Since the beginning of the 21st century, with the discovery of graphene as a milestone, research on low-dimensional quantum materials (including two-dimensional (2D) materials, one-dimensional (1D) nanowires, and zero-dimensional (0D) quantum dots [QDs], etc.) has undergone sustained acceleration.[1–4] Owing to their unique electronic band structures, outstanding mechanical properties, remarkable optical response characteristics, and ultrahigh specific surface area, low-dimensional quantum materials serve as a foundational enabler and a versatile integration platform for next-generation electronic and optoelectronic devices.[5–7]

Traditional semiconductor technologies, particularly silicon-based complementary metal-oxide-semiconductor (CMOS) technology, have been evolving for over half a century in accordance with Moore's Law. They are now increasingly approaching their physical and technological limits.[8,9] This trend manifests in several critical bottlenecks: the intensification of short-channel effects driven by continuous miniaturization of transistors, the difficulty of further reducing static and dynamic power consumption (i.e., the "power wall"), and the significant increase in quantum tunneling current.[10–12] Meanwhile, the evolution of next-generation information technologies has imposed increasingly diverse and stringent performance requirements on semiconductor devices. It spans higher computational throughput per unit energy, ultralow energy consumption per operation, integrated multifunctional capabilities (including in situ sensing, electroluminescence, mechanical flexibility, and heterogeneous integration), and robust adaptability to emerging application paradigms, such as wearable electronics,[13,14] transparent interactive displays,[15,16] and implantable biomedical systems.[17,18] Low-dimensional quantum materials, with their intrinsic atomic-scale thickness, tunable bandgap, ultrahigh carrier mobility, exceptional light absorption and radiative recombination efficiency, as well as outstanding mechanical flexibility and structural deformability, provide a novel material platform and device paradigm for overcoming the aforementioned physical and engineering bottlenecks.

DOI: 10.1201/9781003655305-11

This chapter will systematically explore advances in research on novel electronic and optoelectronic devices based on low-dimensional quantum materials. It will focus on detailing cutting-edge developments, operating principles, and performance advantages across three major domains: low-power electronic devices, optoelectronic devices, and transparent and flexible electronic devices. Finally, it will analyze the primary challenges currently faced and future development directions.

8.2 LOW-POWER ELECTRONIC DEVICES

As Moore's Law approaches its physical and engineering limits, traditional silicon-based CMOS technology faces a series of increasingly severe challenges during its ongoing scaling process. As device feature sizes shrink to the nanometer scale and even sub-5 nm dimensions, the physical behavior of silicon transistors undergoes significant changes: the gate's electrostatic control over the channel deteriorates sharply, leading to increasingly pronounced short-channel effects such as threshold voltage roll-off, reduced leakage potential barrier, and carrier velocity saturation.[11,19,20] The probability of direct tunneling between the source and drain and band-to-band tunneling (BTBT) increases significantly, causing the off-state leakage current to grow exponentially. Meanwhile, the equivalent oxide thickness of the gate dielectric layer is difficult to further reduce under high electric fields, limiting gate control efficiency and exacerbating the dual pressures of dynamic and static power consumption. Furthermore, quantum confinement effects and interfacial roughness scattering are significantly amplified at the atomic scale, leading to increased device parameter dispersion and posing severe challenges to manufacturing yield and long-term reliability. Against this backdrop, how to systematically reduce energy consumption per unit of computation, while maintaining continuous improvements in computational performance, integration density, and functional scalability, has become a core scientific challenge and a key technological proposition jointly addressed by the global semiconductor industry and academia. Owing to their distinctive optical response characteristics, superior charge carrier transport properties, highly programmable heterojunction interfaces, and tunable bandgap architectures, low-dimensional quantum materials have emerged as a leading class of channel materials capable of overcoming current energy-efficiency limitations and enabling the development of next-generation ultralow-power electronic devices.

Tunneling field-effect transistors (TFETs) constitute an emerging class of semiconductor field-effect devices. While their structure resembles that of metal-oxide-semiconductor field-effect transistors (MOSFETs), their switching mechanism differs fundamentally: TFETs achieve on/off switching by regulating the probability of BTBT at the source-channel interface through gate voltage control. This mechanism overcomes the thermodynamic subthreshold swing (SS) limitations of conventional MOSFETs, theoretically enabling SS below 60 mV/dec.[21,22] Consequently, it holds significant potential for drastically reducing static power consumption. Although its on-state current density currently remains lower than that of advanced MOSFETs, it demonstrates substantial value in ultralow-power integrated circuit applications.[23,24]

Figure 8.1 illustrates an N-type TFET. The source and drain terminals are implemented using heavily doped p-type and n-type semiconductor regions, respectively.

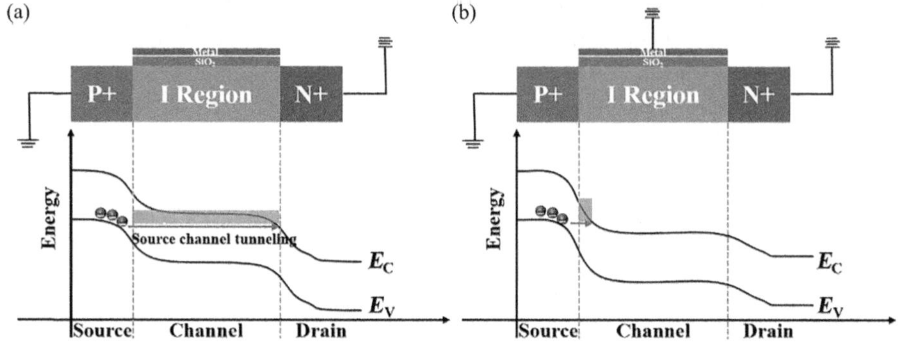

FIGURE 8.1 BTBT mechanism in an n-type TFET. (a) Off state. (b) On state.

The gate is separated by an insulating layer (silicon dioxide, SiO_2) and connected to the tunnel region, which is typically made of intrinsic semiconductor silicon. First, the operating principle of this device is described. In the off state, the source terminal is grounded (or biased at a low potential), whereas the drain is applied with a forward bias (high level). The corresponding energy band distribution across different regions of the device is illustrated in the band diagram shown in Figure 8.1a, where E_c denotes the conduction band and E_v denotes the valence band. For the device to conduct, valence band electrons in the source region (a heavily doped p-type region) must traverse the tunneling barrier at the source-channel interface. This process is termed source-channel BTBT, which falls under the quantum mechanical tunneling effect. When the tunneling region is relatively wide, the left P-region can be approximated as an electron emission source, while the right intrinsic channel forms a wide barrier region. At this point, the electron tunneling probability is significantly reduced, and most electrons will be reflected back to the source region by the barrier. Upon application of a positive gate voltage (V_{GS}), the device transitions from the off state to the on state. As shown in Figure 8.1b, gate voltage modulation intensifies the band bending at the source-channel interface, significantly reducing the barrier width and steepening its profile. Under narrow tunneling conditions, the left-side p-type source region acts as a carrier emission source, while the right-side intrinsic channel forms a narrow barrier region. This substantially increases the electron tunneling probability, thereby achieving a significant conduction current. This gate-voltage-dependent barrier regulation mechanism constitutes the core switching characteristic of TFETs.

Several theoretical models have been established to compute the quantum mechanical tunneling probability through triangular potential barriers; comprehensive reviews of these approaches are available in the literature.[25,26] Among them, the Wenzel–Kramers–Brillouin (WKB) approximation is the most widely employed analytical method for evaluating BTBT probability in TFETs. The probability of valence-band electrons tunneling from the p-type source region to the intrinsic channel in TFETs (T_{WKB}) can be solved using the WKB approximation[23,27,28]:

$$T_{\mathrm{WKB}} \approx \exp\left(\frac{-4\lambda E_g^{3/2}\sqrt{2m_t^*}}{3q\hbar\left(E_g + \Delta\varphi\right)}\right)$$

where m_t^* denotes the tunneling effective mass, E_g represents the bandgap energy, $\hbar$ is the reduced Planck constant, q is the elementary charge, $\Delta\varphi$ signifies the energy difference between the valence band maximum (VBM) of the source region and the conduction band minimum (CBM) of the channel (i.e., the tunneling energy window), and λ characterizes the effective shielding length, quantifying the spatial decay scale of band bending at the tunneling junction. In a TFET, under constant drain voltage (V_{DS}), an increase in V_{GS} reduces the effective shielding length λ and enlarges the energy difference $\Delta\varphi$ between the source CBM and the channel VBM. As a result, the drain current (I_{DS}) exhibits a super-exponential dependence on V_{GS}. Accordingly, the SS of TFETs is no longer constant but exhibits strong gate voltage dependence. To achieve high on-state current, key material and structural parameters must be co-optimized: specifically, reducing the tunneling effective mass m_t^*, the bandgap energy E_g, and the effective shielding length λ, thereby increasing the BTBT probability T_{WKB}.

The core objective of high-performance TFET development is to synergistically optimize turn-on current and SS across a broad subthreshold operating range (spanning multiple orders of magnitude in current dynamic range), achieving an optimal trade-off between the two. Enhancing the on-state current of TFETs critically depends on maximizing the BTBT probability under low gate bias conditions. This objective can be achieved by reducing effective mass m_t^*, the bandgap energy E_g, and the effective shielding length λ. However, there exists an inherent physical conflict between minimizing SS and enhancing on-state current in TFETs, arising from the strong coupling mechanism among band engineering, electrostatic control, and quantum tunneling. Specifically, although reducing the bandgap and effective tunneling carrier mass enhances the tunneling probability T_{WKB}, it also accelerates the rate of shift between the CBM and VBM at the source/channel interface with respect to gate voltage, thereby degrading the SS. Similarly, while reducing the shielding length λ enhances gate electrostatic control and shortens the tunneling path, excessive band bending amplifies the electrostatic shielding effect. This limits the tunable range of the tunneling width with respect to the tunneling energy window, ultimately degrading the SS due to a weakened modulation response of the gate voltage to the tunneling energy window. Therefore, the interdependent effects of material parameters (E_g, m_t^*) and electrostatic parameters (λ, oxide thickness, channel doping) on open-circuit current and SS necessitate their synergistic optimization through systematic co-design of the material system and device architecture.

Breakthroughs in 2D materials such as graphene,[29,30] transition metal dichalcogenides (TMDCs),[31,32] and black phosphorus (BP)[33,34] have opened new avenues for the design of TFETs. These materials exhibit distinct advantages arising from their atomic-scale thickness, dangling-bond–free planar surfaces, and layer-dependent electronic structures. In atomically thin 2D semiconductors, intrinsic quantum confinement effects localize charge carriers within sub-nanometer dimensions, thereby enabling gate voltage to exert global electrostatic control over the channel potential. The bandgap tunability of 2D affords a multidimensional design space for heterojunction band engineering. Channel thicknesses achieved with atomic-level precision suppress interlayer uniformity fluctuations, thereby promoting steep transitions at the band edges and optimizing interband tunneling properties.

Crucially, 2D materials possess an extremely short electrostatic shielding length (λ). This characteristic effectively compresses the tunneling potential barrier width and enhances the probability of interband tunneling, forming the physical basis for achieving steep SS. For example, previous studies have demonstrated that TFETs in which the conventional silicon material in the channel is replaced with atomically thin MoS_2 and highly doped germanium (Ge) as the source electrode exhibit excellent steep turn-on characteristics, with an average SS as low as 31.1 mV/dec.[22] More recently, a study reported a BP-based natural heterojunction TFET that utilizes the spatial gradient in BP layer thickness to form the heterojunction, thereby fundamentally avoiding interface issues and achieving a SS of 22.9 mV/dec across a current dynamic range spanning four orders of magnitude.[38] In addition, the design of the gate structure is crucial for the precise control of the band structure in TFETs and is a key factor in improving the device's switching characteristics. Wang et al. exploited the inherently strong ferroelectric field of 2D $CuInP_2S_6$ (CIPS) to implement a ferroelectric gate-controlled 2D TFET, in which the channel was realized either as a tungsten diselenide (WSe_2)/molybdenum disulfide (MoS_2) heterojunction or a WSe_2 homostructure, as shown in Figure 8.2a. The device demonstrated an ultralow SS of 14.2 mV/dec at room temperature and an on/off current ratio exceeding 10^8, as presented in Figure 8.2b and c.[35] Huo et al. developed a novel ultra-thin, short-channel tunneling transistor that departs from conventional 2D planar architectures.

As illustrated in Figure 8.2d, the device employs a vertically stacked MoS_2/WSe_2 heterojunction that serves as the gate-controlled tunnel barrier, while an in-plane–oriented WSe_2 monolayer functions as the conduction channel. Switching operation is achieved through dynamic regulation of the barrier height by the heterojunction gate, enabling reversible switching between thermal electron emission and BTBT transport mechanisms.[36] Figure 8.2f presents a schematic diagram of the energy band and the electron transport process under different top-gate voltages and a fixed drain voltage. When $V_{TG} < 0$ V, holes accumulate in the WSe_2 channel, and the double junction forms an n^-–p–n^- structure; at this point, the barrier height of the MoS_2/WSe_2 heterojunction at the source significantly increases. Under an applied drain-source voltage, electrons in the MoS_2 source region must overcome this high potential barrier and traverse the p-type WSe_2 channel via thermal emission, generating only a negligible thermal electron current and maintaining the device in a deep-off state. As the gate voltage gradually approaches 0 V, the degree of band bending gradually decreases and begins to reverse; the potential barrier drops to the level of the intrinsic conduction band edge, and the device forms an n^--i-n^- structure, generating a certain amount of drain-source current. Conversely, when $V_{TG} > 0$ V, the WSe_2 channel transitions to electron accumulation, and the transport channel changes to an n^-–n–n^- structure; as the top-gate voltage increases further, a steep local band peak

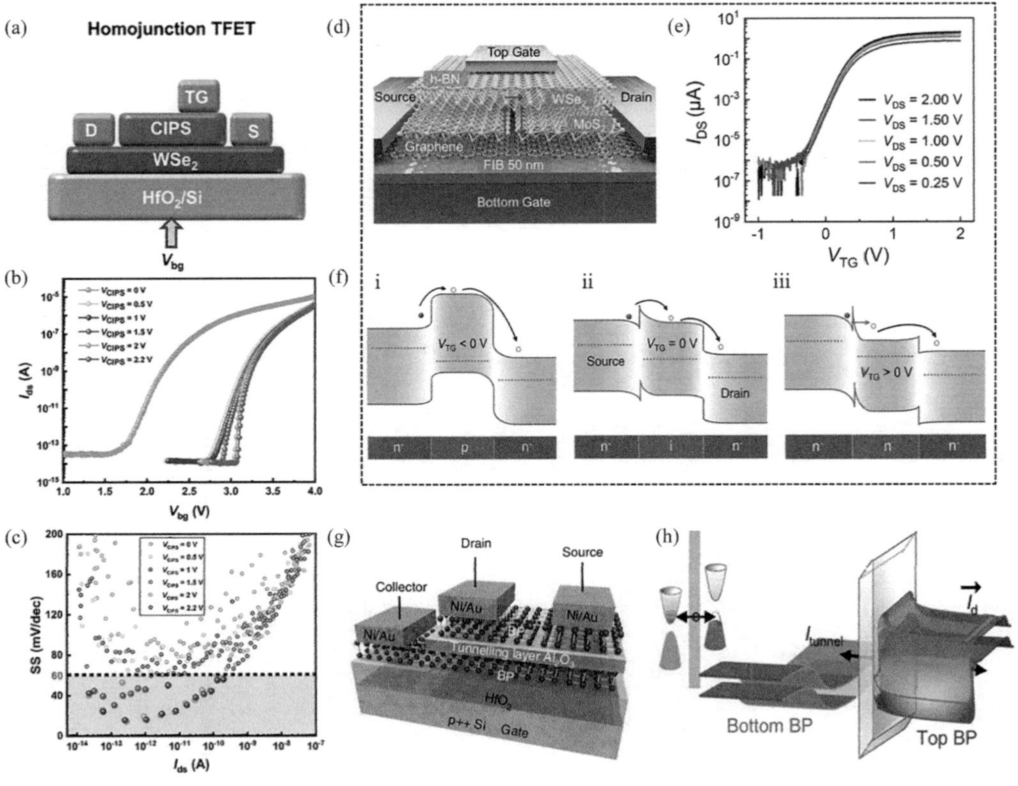

FIGURE 8.2 (a) Schematic cross-section of WSe_2 homojunction TFET.[35] (b) Transfer characteristic curves with V_{CIPS} changing from 0 to 2.2 V at $V_{DS} = 1$ V.[35] (c) I_{DS} versus SS data extracted from the transfer curves shown in (b).[35] (d) Schematic diagram of the short-channel tunneling transistor.[36] (e) Transfer characteristic curves of the device at different V_{DS}.[36] (f) Schematic diagram of the energy band and the electron transport process under different top-gate voltages and fixed drain voltage.[36] (g) Schematic diagram of the structure of a 2D BP/Al_2O_3/BP van der Waals heterojunction TFET.[37] (h) 2D energy band diagram of the 2D BP/Al_2O_3/BP TFET.[37]

forms at the conduction band edge of the MoS_2/WSe_2 heterojunction at the source, significantly narrowing the barrier width and thereby triggering a strong quantum tunneling effect. At this point, the carrier transport mechanism shifts from being dominated by thermal emission to being dominated by interband tunneling, and the steep band peak significantly increases the tunneling probability, enabling the device to achieve high-efficiency conduction.[36] Wu et al. reported a transverse tunneling field-effect transistor based on a 2D BP/alumina/black phosphorus ($BP/Al_2O_3/BP$) van der Waals heterojunction, as illustrated in Figure 8.2g. In contrast to conventional tunneling diodes, the tunneling carriers in this device move in the lateral direction of the driving current. The electrostatic effect generated by these carriers significantly modulates the channel carrier density, thereby achieving a substantial negative differential resistance (NDR) characteristic, as shown in Figure 8.2h.[37]

8.3 OPTOELECTRONIC DEVICES

Since the beginning of the 21st century, optoelectronics has continued to deepen its integration with cutting-edge disciplines such as quantum optics and quantum communication. In particular, quantum optics research—which focuses on the precise manipulation of photonic quantum states—has established foundational theoretical frameworks and enabled critical enabling technologies for strategic domains such as quantum computing and quantum communication. At the same time, the rapid advancement of nanotechnology has significantly expanded the scope of research and the boundaries of application in optoelectronics; nanophotonics and nano-optoelectronic devices have become a key focus of both the international academic community and industry. As the core functional component of optoelectronic systems, photodetectors play an indispensable role in the detection, conversion, and transmission of optical information, and have been widely adopted in numerous critical fields, including high-resolution imaging, high-speed optical communications, and precision optical metrology. The trend toward intelligent optical information processing is driving the evolution of photodetectors toward multispectral responsivity, ultralow power consumption, and neuromorphic sensing functionality. Traditional silicon-based detectors, constrained by their intrinsic bandgap of 1.12 eV, exhibit weak response in the ultraviolet (UV) band (< 400 nm) and the near-infrared (NIR) band (>1,100 nm).[39, 40] Furthermore, their rigid substrate structure makes it difficult to meet the integration requirements of flexible wearable electronic systems.

According to peer-reviewed studies, the external quantum efficiency (EQE) of conventional silicon-based photodetectors at the 1,550 nm telecommunications wavelength is typically below 20%, underscoring a fundamental performance limitation for optical communication and photonic sensing applications. In contrast, low-dimensional materials, leveraging intrinsic quantum effects induced by dimensional confinement, offer a novel technical approach to overcoming these limitations. Monolayer MoS_2 exhibits a direct bandgap of 1.8 eV and an absorption coefficient exceeding 10^7 m^{-1}, which is four orders of magnitude higher than that of bulk silicon. BP, due to its thickness-dependent tunable bandgap (0.3–2.0 eV), can effectively cover the mid-infrared atmospheric transparency window (3–5 μm). Meanwhile, lead sulfide QDs achieve high EQE in the short-wave infrared (SWIR) band (1.3–1.5 μm) via precise nanoscale size engineering. This paradigm shift, replacing traditional material modification with dimensional tuning, is systematically driving the coordinated development of photodetectors toward ultrahigh sensitivity, broad spectral response, and flexible heterogeneous integration.

Notably, 2D layered materials confine carrier transport to a single planar dimension at the nanoscale; intralayer atomic bonding is mediated by strong covalent interactions, whereas interlayer stacking is governed by weak van der Waals forces, resulting in crystalline membranes of atomic thickness. Representative examples include graphene, TMDCs, and BP. In contrast to conventional epitaxial semiconductors, these 2D materials can be fabricated at low cost and on a large scale using both top–down and bottom–up approaches, and they exhibit exceptional optical transparency coupled with broadband (UV-to-mid-infrared) absorption, thereby offering broad application prospects in high-performance photodetectors. Xia et al. demonstrated a BP mid-infrared photodetector operating at a wavelength of 3.39 μm, which achieves high intrinsic gain and an external responsivity of 82A W^{-1}, as shown in Figure 8.3a.[41] Wang et al. innovatively designed and fabricated an ultra-broadband multilayer MoS_2 photodetector by controlling the deviation from the perfection of the atomic lattice. As shown in Figure 8.3b, the device exhibits continuous and stable photoresponse characteristics across a broad spectral range from 445 nm (blue light) to 2,717 nm (mid-infrared), achieving a total coverage of 2,272 nm. Its responsivity reaches up to 50.7 $mA{\cdot}W^{-1}$, and the specific detectivity reaches 1.55×10^9 Jones.[42]

Notably, 1D nanostructured materials (such as nanowires, nanorods, and nanotubes) strictly confine carrier motion to a 2D plane at the nanoscale, thereby inducing significant quantum confinement effects and novel physical properties; their high specific surface area and size-dependent effects also confer unique advantages in optoelectronic devices. Li et al. have developed an ultra-broadband self-powered photodetector based on low-dimensional $Nb_3Se_{12}I$ nanowires, integrating broadband photoresponse with biomimetic visual functionality. As shown in Figure 8.3c, it can mimic the image adaptation process of human vision under varying lighting conditions.[43] The team led by Christoph J. Brabec proposed a Dynamic Template-Assisted Coating strategy to fabricate large-area, 1D perovskite single-crystalline nanowire arrays ($MAPbBr_3$ SCNWAs). Photodetectors based on these arrays demonstrate outstanding room-temperature photoelectrical performance: a high responsivity of 1,660 $A{\cdot}W^{-1}$, a specific detectivity of 3.9×10^{14} Jones, and a linear dynamic range of 160.3 dB, while also featuring high response speeds and excellent environmental stability, demonstrating application potential in flexible, high-sensitivity photodetection and wearable sensing systems.[45] Notably, 0D nanostructured materials (such as nanoparticles and QDs) confine the motion of excitons in three spatial dimensions. When the particle size is smaller than the de Broglie wavelength, quantum confinement effects occur; by controlling the particle size and morphology, the energy levels can be modified, thereby increasing the density of states. As a result, 0D materials possess unique and exceptional photosensitive properties, along with three-dimensional (3D) tunability, enabling the development of more compact and multifunctional optoelectronic detection systems by reducing device size and improving sensor resolution. Lee et al. reported a pioneering architecture for a CQD-based infrared photodetector that uses kinetically pumped avalanche multiplication. This design achieves kinetic-pumped charge multiplication by

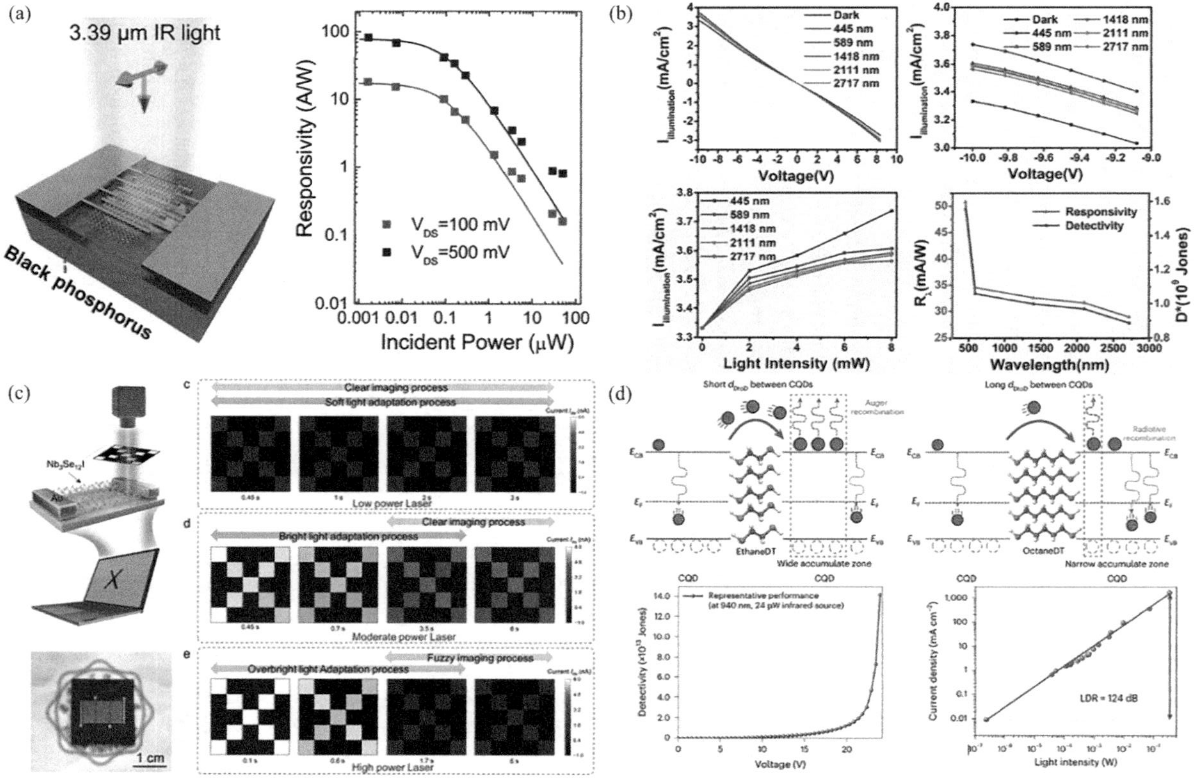

FIGURE 8.3 (a) Schematic diagram of a BP mid-infrared photodetector and power dependence of the responsivity.[41] (b) Photoresponse performance of ultrabroadband MoS_2 photodetectors.[42] (c) Bionic visual adaptation capabilities of ultrawideband self-powered photodetectors based on low-dimensional $Nb_3Se_{12}I$ nanowires.[43] (d) Schematic illustration of the mechanism whereby the d_{DtoD} parameter modulates inter-dot charge transport and hot-carrier multiplication kinetics in colloidal quantum dots (CQDs), and its impact on the photoresponse performance of CQD-based infrared photodetectors under 940 nm illumination.[44]

applying a strong electric field to a CQD layer with a thickness exceeding 540 nm, thereby imparting sufficient kinetic energy to the photogenerated electrons to overcome the material's bandgap. By precisely controlling the average spacing between adjacent QDs to approximately 4.1 nm, a balance can be achieved between the effects on ionization and electronic transitions, thereby significantly improving performance. The optimized device achieves a specific detectivity of 1.4×10^{14} Jones at 940 nm and an avalanche multiplication factor of up to 85, as shown in Figure 8.3d.[44]

However, photodetectors still face multiple challenges in actual fabrication and application: issues such as narrow light absorption spectra, short carrier recombination lifetimes, high contact resistance at the electrode/active layer interface, and insufficient environmental stability in single low-dimensional materials severely limit improvements in device performance. Therefore, through rational device design, such as optimizing material selection, interfacial properties, structural architecture, and fabrication processes, to achieve a precise alignment between the intrinsic properties of the materials and the functional requirements of the devices has become the key to overcoming the performance bottlenecks of low-dimensional material photodetectors. Currently, performance advances in low-dimensional photodetectors are predominantly enabled by design strategies: van der Waals heterostructure integration, interfacial engineering, and doping. By optimizing the entire process of photogenerated carrier generation, separation, transport, and collection, these strategies effectively suppress non-radiative recombination and interface scattering losses, thereby overcoming the physical limitations of single-material systems and comprehensively enhancing the device's sensitivity, response speed, and stability.

By constructing heterostructures and controllably combining low-dimensional materials with different dimensions or physical properties, it is possible to achieve synergistic complementarity among the advantageous characteristics of each component, thereby effectively expanding the functional boundaries and performance limits of devices. Based on the dimensional relationships between their constituent components, heterostructures can be classified into two categories: same-dimensional heterostructures (such as 2D MoS_2/WS_2 van der Waals heterojunctions) and mixed-dimensional heterostructures (such as 0D QDs/2D MoS_2 and 1D nanowires/2D graphene heterostructures). The key advantage of same-dimensional heterostructures lies in their ability to precisely control the band structure, thereby effectively promoting the efficient separation and directed transport of photogenerated carriers. Based on the energy level arrangement at the interface, these structures primarily exhibit three typical band configurations: Type I, Type II, and Type III.

The following are typical examples of these three band configurations. Lv et al. employed a two-step CVD process to synthesize atomical WS_2/SnS_2 vertical heterojunctions. This heterostructure features a prototypical Type-II band alignment. Under illumination, the built-in electric field at the heterojunction interface rapidly drives the spatial separation of photogenerated carriers: electrons spontaneously flow toward SnS_2, while holes remain in WS_2. This highly efficient charge separation mechanism extends the carrier lifetime, resulting in substantial photoelectric gain. Under white-light illumination, the device achieves a specific detectivity of 5.8×10^{10} Jones and a high responsivity of 98.33 A W^{-1}, placing it at the forefront of its class, as shown in Figure 8.4a.[48] Gong et al. reported a α-In_2Se_3/$PdSe_2$ ferroelectric van der Waals heterojunction photodetector, achieving both high responsivity and broadband spectral response, as summarized in Figure 8.4b. Figure 8.4c presents a schematic diagram of the band structure of the α-In_2Se_3/$PdSe_2$ van der Waals heterojunction device before and after metal electrode contact. Prior to contact, the electron affinities of α-In_2Se_3 and $PdSe_2$ nanosheets are −3.60 eV and −4.35 eV, respectively, corresponding to optical bandgaps of 1.43 (57 nm) and 0.34 eV (10.5 nm), respectively. Upon contact, an intrinsic built-in electric field (E_{bi}) pointing from α-In_2Se_3 toward $PdSe_2$ forms at the heterojunction interface, and positive and negative charge centers accumulate on either side of the α-In_2Se_3 to $PdSe_2$ interface, respectively. Consequently, the band edges of α-In_2Se_3 undergo upward bending, while those of $PdSe_2$ exhibit downward bending. Owing to the isotropic nature of the α-In_2Se_3/$PdSe_2$ heterojunction, a depletion region forms on the α-In_2Se_3 side near the heterojunction interface, thereby creating a bandgap-aligned (Type I) heterojunction. In a unidirectional depletion heterojunction, only one type of photogenerated carrier (electrons or holes) in the depletion region crosses the heterojunction interface and rapidly recombines with the opposite carrier in the accumulation region. Consequently, interfacial recombination of electron–hole pairs is suppressed, and carrier collection efficiency is enhanced, facilitating ultrafast and highly sensitive photodetection.[46] Li et al. fabricated vertically stacked type-III GeSe/SnS_2 heterojunctions, and the corresponding photodetectors demonstrated optimal photoresponse performance at 532 nm excitation. Under external illumination, the built-in electric field induced by band offsets at the heterojunction drives highly efficient spatial separation of photogenerated electron–hole pairs: electrons selectively migrate to the SnS_2 layer, while holes are directionally transported to the GeSe layer, as illustrated in Figure 8.4d. At low light power densities, the device achieves a specific detectivity of 7.3×10^{11} Jones and an EQE of 30.3%, as presented in Figure 8.4e.[47]

Mixed-dimensional heterostructures enable substantial enhancement of optoelectronic device performance through the rational design and synergistic coupling of the intrinsic physical properties of 0D, 1D, and 2D materials. As prototypical 0D semiconductor nanomaterials, CQDs have emerged as critical photosensitive components in high-performance photodetectors, owing to their excellent solution processability, size-dependent bandgap tunability, and strong light absorption capabilities. For instance, Jong-Soo Lee's team reported a 0D/2D heterostructure composed of 0D Ag_2Te CQDs and 2D MoS_2 nanosheets, as shown in Figure 8.5a. By utilizing an interface spacing (<2–3 nm) shorter than the exciton diffusion length of Ag_2Te (~7.5 nm), the heterostructure enables efficient photogenerated electron transfer from the QD to MoS_2 and concurrent hole localization within the QDs, thereby creating a "photo-doping effect." This boosted the optical gain from the low values typically observed in conventional QD films to 7.8 times, enabling high-sensitivity detection in the SWI (SWIR) band.

As shown in Figure 8.5b and c, based on this hybrid structure, a 32 × 32 pixel array SWIR imaging device was successfully fabricated, enabling spatial resolution pattern reconstruction and real-time imaging.[49] In addition to 0D materials, 1D quantum structures represent promising candidates for high-performance optoelectronic devices, owing to their quantum confinement-induced high carrier density and

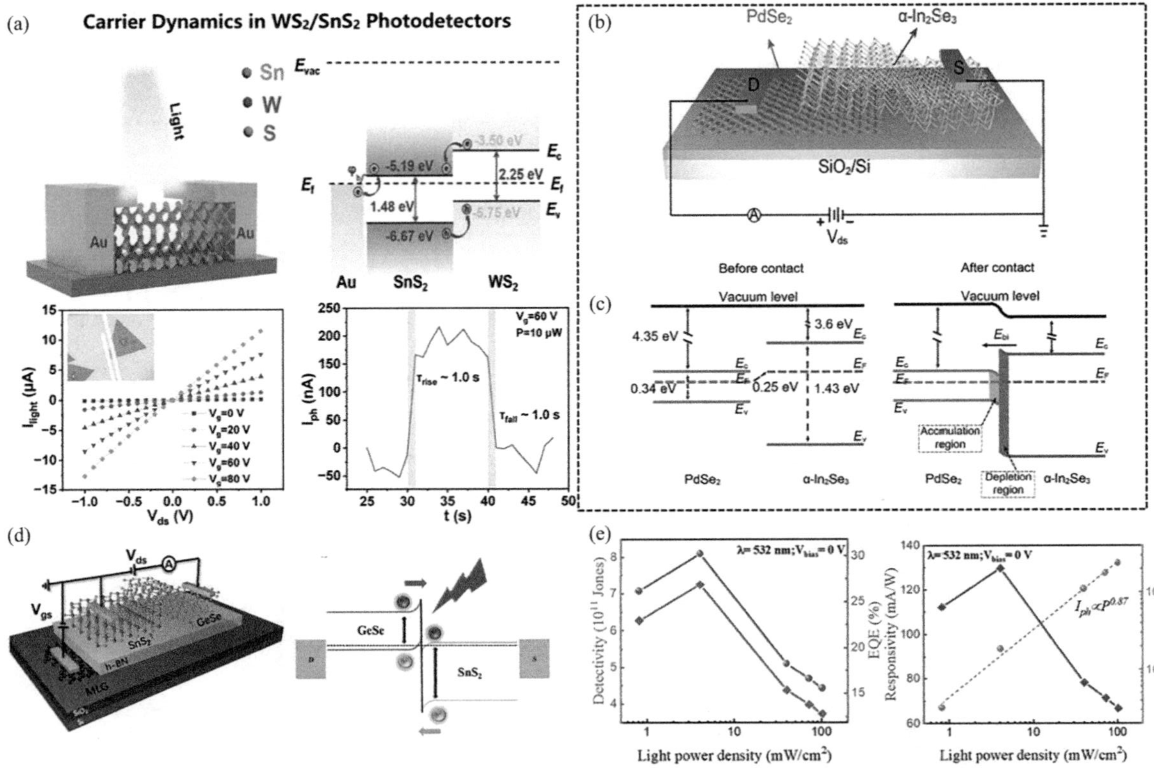

FIGURE 8.4 (a) Schematic illustration and optoelectronic performance of the WS_2/SnS_2 heterostructure-based photodetector. Band alignment diagram and carrier transfer mechanisms in the $Au/SnS_2/WS_2$ heterostructure. (b) Schematic illustration of the α-$In_2Se_3/PdSe_2$ photodetector.[46] (c) Schematic diagram of the energy bands of an α-$In_2Se_3/PdSe_2$ van der Waals heterojunction photodetector before and after contact.[36] (d) 3D Schematic image of the $GeSe/SnS_2$ heterostructure on MLG/h-BN structure on SiO_2/Si substrate; schematic energy band diagram illustrating photoexcitation-driven electron–hole separation.[47] (e) Photodetection performance based on $GeSe/SnS_2$ photodetectors.[47]

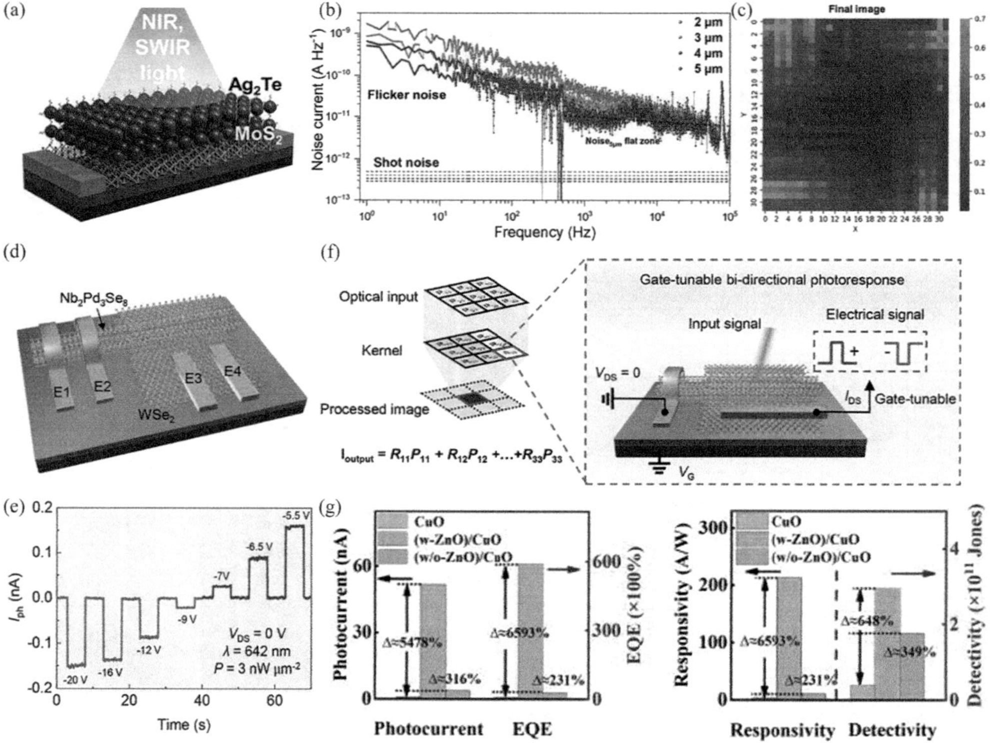

FIGURE 8.5 (a) Schematic illustration of the 0D/2D hybrid photodetector based on the Ag_2Te CQDs–MoS_2 nanosheet heterostructure.[49] (b) Frequency-dependent noise current of Ag_2Te/MoS_2 hybrid photodetectors with different channel lengths (2–5 μm), exhibiting characteristic flicker (1/f) behavior.[49] (c) Final image synthesized from the four angular (0°, 90°, 180°, and 270°) response maps.[49] (d) Schematic illustration of a self-powered photodetector based on a van der Waals heterojunction formed between 1D $Nb_2Pd_3Se_8$ and 2D WSe_2.[50] (e) Gate-tunable Negative/positive photocurrent switching illumination at 642 nm with P = 3 nW·um^{-2}.[50] (f) Schematic of in-sensor image processing with the kernel pixels utilizes the gate-tunable negative and positive photocurrent switching in $Nb_2Pd_3Se_8/WSe_2$ heterostructure device.[50] (g) Optoelectronic performance of 0D/1D ZnO/CuO heterojunction photodetector.[51]

directional charge transport characteristics. The rational integration of 1D/2D mixed-dimensional heterostructures enables synergistic performance enhancement, establishing a robust foundation for next-generation optoelectronic devices. For example, Ji-Hee Kim et al. reported a self-powered photodetector based on a van der Waals heterojunction formed between 1D $Nb_2Pd_3Se_8$ and 2D WSe_2 (Figure 8.5d). Under 375 nm UV illumination, this device exhibits excellent high-energy photon conversion efficiency, delivering a responsivity of 232 mA W^{-1}, an EQE of 77%, rise/fall times of approximately 3 us, and a specific detectivity of 6.35×10^{10} Jones, with a broad spectral response range spanning from the UV to the NIR. As shown in Figure 8.5e, gate modulation of the built-in electric field enables reversible switching between n–p and n–n^+ junction configurations, thereby achieving precise reversal of the photogenerated carrier drift direction and bidirectional (positive/negative) photocurrent output. This gate-tunable bidirectional photocurrent response has been further leveraged for analog-domain convolutional image processing, as shown in Figure 8.5f.[50] In addition, in the field of low-dimensional hybrid optoelectronic detection, 0D/1D heterojunctions have been extensively studied because they combine the synergistic effects of different dimensions, thereby improving carrier transport efficiency.[52–54] For example, 0D/1D ZnO/CuO heterojunction photodetector exhibits high photoconversion efficiency, as shown in Figure 8.5g.[51]

Interface properties constitute a critical determinant of the performance of photodetectors based on low-dimensional materials; specifically, the quality of interfacial contact between the low-dimensional material and both the electrodes and substrate directly determines the efficiency of carrier injection and transport losses. When a metal electrode contacts a low-dimensional material, conventional fabrication techniques (such as thermal metal evaporation) frequently introduce interface trap states at the contact surface, resulting in severe Fermi-level pinning, an elevated Schottky barrier height, substantially increased contact resistance, and hindered charge transport efficiency. Therefore, the primary objective of interface engineering is to establish high-quality interfaces that minimize contact resistance and suppress non-radiative carrier recombination. Current interface engineering strategies primarily encompass electrode contact engineering and interfacial passivation. Regarding electrode contact optimization, a research team at Tsinghua University recently replaced the conventional high-temperature thermal evaporation process with a physical transfer technology. By utilizing van der Waals forces to bond gold electrodes to WS_2/SnS_2 heterojunctions, the researchers avoided damage to the material structure caused by atomic bombardment. This strategy successfully reduced the Schottky barrier height at the contact interface from 110 to 59 meV, increasing the device current by three orders of magnitude and significantly improving carrier collection efficiency.[48]

Other studies have demonstrated an interface engineering strategy that incorporates graphene as a contact electrode in van der Waals heterostructures composed of violet phosphorus and MoS_2. As a high-quality 2D conductive electrode, graphene can effectively reduce the contact potential barrier and facilitate interfacial charge transfer, thereby significantly enhancing the device's photoconversion efficiency and carrier extraction capability.[55] In addition, the researchers designed a $Cr/WSe_2/Au$-based bottom-contact Schottky photodetector that achieves low-damage interface contact through the bottom-embedded electrode and the top Se buffer layer, resulting in an ultrafast optical response and excellent photodetection performance.[56] In terms of surface passivation, physical and chemical passivation strategies, such as coating or encapsulation of low-dimensional material surfaces, effectively passivate surface defect states and suppress non-radiative carrier recombination, thereby significantly improving the sensitivity, response speed, and long-term operational stability of photodetectors. Commonly used passivation materials include aluminum oxide (Al_2O_3)[57, 58] and SiO_2.[59] For example, depositing an ultrathin Al_2O_3 passivation layer on the surface of a $SnSe_2$/SnSe heterojunction using atomic layer deposition not only prevents oxidation and chemical corrosion of the heterojunction surface but also further enhances the stability of the $SnSe_2$/SnSe photodetector.[60] In addition, integration of an ultrathin Al_2O_3 passivation layer into the MoS_2/Al_2O_3/Si flexible heterostructure yields outstanding optoelectronic performance: a responsivity of 0.047 A W^{-1}, a specific detectivity of 5.67×10^{11} Jones, and rise/fall times of 25.6 and 3.2 μs, respectively.[61]

Doping is one of the effective strategies for modulating the optoelectronic properties of low-dimensional materials. Introducing donor- or acceptor-type dopants into low-dimensional materials enables precise control over carrier concentration and band structure, thereby enhancing carrier mobility, optical absorption coefficients, and carrier separation efficiency driven by an intrinsic electric field, ultimately optimizing the device's response performance. The key to doping modification lies in selecting the appropriate dopant elements and doping methodologies to avoid introducing excessive defects and ensure lattice continuity and interface quality. For example, by introducing oxygen doping onto the surface of a $ZrTe_3$ nanoribbon using a 355 nm UV laser, an ultra-broadband polarization detector with a back-to-back Schottky barrier structure was successfully fabricated. Under zero-bias, room-temperature operation and in the absence of antenna integration, this device exhibits continuous polarization response across the 355 nm to 1.43 mm wavelength range (UV to terahertz).[62] Al doping into colloidal $CuInS_2$ QDs induced the growth of a surface self-oxidation protective layer, which significantly suppressed non-radiative recombination in the QDs, optimized optical properties, and enhanced photochemical stability.[63]

Doping concentration is one of the critical parameters for improving the performance of photodetectors. Taking the Mn-doped ZnSe microsphere system as an example, when the Mn doping molar ratio is 2%, the carrier density of the material increases significantly, and the absorption coefficient in the UV–Vis region is markedly enhanced.[64] In Mg-doped Ga_2O_3 photodetectors, the dark current decreases with increasing Mg doping concentration, whereas the photocurrent exhibits a non-monotonic dependence, peaking at 4.2% Mg doping concentration. The underlying physical mechanism can be attributed to the following two factors: On the one hand, Mg doping increases the optical bandgap of the film, resulting in a higher metal-semiconductor contact barrier height, which effectively suppresses the progress of carrier transport and thereby reduces the dark current. On the other hand, the acceptor energy levels introduced by Mg doping enhance the concentration and separation efficiency of photogenerated carriers, thereby increasing the photocurrent. However, when the doping concentration is too high, the excessively wide bandgap of the film requires a higher excitation energy to generate photogenerated electron–hole pairs;

consequently, a decrease in photocurrent is observed under such conditions.[65] Therefore, defect engineering necessitates precise control of defect concentration. Introducing deep-level defects as carrier traps effectively suppresses non-radiative recombination while simultaneously modulating band structure to enhance light absorption efficiency. Excessive defect concentration can lead to increased carrier scattering, which in turn degrades device performance.

8.4 TRANSPARENT AND FLEXIBLE ELECTRONIC DEVICES

Driven by sustained advances in low-dimensional quantum materials research, optoelectronic devices are undergoing a paradigmatic shift toward transparency, mechanical flexibility, and structural miniaturization. As a paradigmatic embodiment of the convergence between low-dimensional quantum materials and optoelectronic technology, transparent flexible optoelectronic devices transcend the intrinsic constraints of conventional rigid optoelectronic devices in terms of structural form and functional integration, achieving the synergistic optimization of optical performance, mechanical flexibility, and optical transparency. These devices have demonstrated broad application prospects in fields such as wearable sensing, electronic skin, flexible displays, and optoelectronic energy harvesting. This section will systematically examine the fundamental design principles, key fabrication methodologies, archetypal device architectures, and representative applications of transparent flexible optoelectronic devices, providing a foundational reference for future research and applications in this field.

The key to fabricating transparent and flexible optoelectronic devices lies in the precise deposition and patterning of low-dimensional quantum materials on flexible substrates, while ensuring the synergistic optimization and functional integration of optical transparency, mechanical flexibility, and optoelectronic conversion efficiency. The solution-based method is currently the most promising technology for the large-scale fabrication of transparent and flexible optoelectronic devices, owing to its cost-effectiveness, process simplicity, capability for uniform, large-area thin-film deposition, and excellent compatibility with low-temperature manufacturing protocols.[66–69] As a typical solution-based processing technique, inkjet printing enables the deposition of various transparent conductive inks, including polymers, carbon-based materials, and metal nanowires, onto a wide range of substrates. It offers the advantages of precise patterning and high material utilization, allowing for the direct fabrication of high-resolution functional structures (such as flexible electrode arrays and light-emitting pixel units) without the need for additional etching processes. As shown in Figure 8.6a, electronic textiles with circuit functionality were successfully fabricated by sequentially printing carbon nanotube ink and conductive silver ink onto a cotton fabric substrate.[70] Lithography enables high-precision patterning and is extensively employed in the fabrication of high-performance and nanoelectronic devices, including flexible display pixels, wearable microsensors, and stretchable circuit interconnects. Based on the type of exposure source, lithography technologies are primarily divided into two categories: optical lithography and particle beam lithography. Currently, optical lithography is the most widely used lithography technology and will remain the mainstream approach for the foreseeable future.

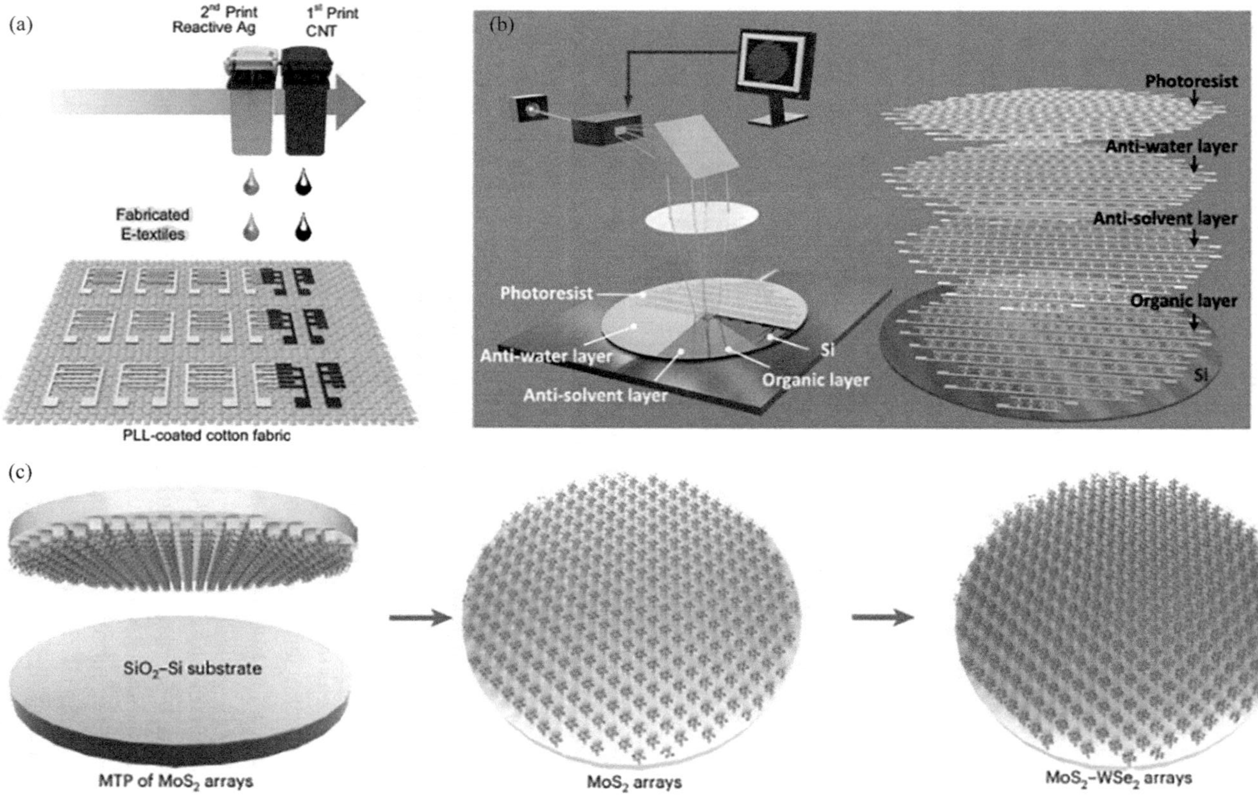

FIGURE 8.6 (a) Inkjet printing for the fabrication of electronic textiles.[70] (b) Schematic illustration of universal microlithographic organic materials.[71] (c) Schematic illustration of wafer-scale 2D MoS_2 array transfer via nanoimprinting.[72]

In particle beam lithography, electron beam lithography, and ion beam lithography technologies allow patterns to be written directly onto silicon wafers without the need for masks. Since increasing the energy of electron and ion beams shortens their wavelengths, the resulting patterns have extremely high resolution. However, the solvents used in conventional photolithography processes can cause organic films to swell or even dissolve, resulting in a significant decline in device performance. Consequently, there is an inherent incompatibility between organic functional materials and conventional photolithography chemical environments, which severely hampers the integration and industrialization of organic and flexible electronic devices.

Recent studies have reported a versatile and non-destructive ultra-thin orthogonal double-protection-layer photolithography method that enables high-precision, non-destructive patterning of organic films, as shown in Figure 8.6b. By introducing an antisolvent protective layer and a hydrophobic protective layer, this method effectively shields organic materials from solvent erosion during photolithography process. The antisolvent protective layer prevents organic solvents used in processes such as photoresist coating and stripping from damaging the underlying organic film. The hydrophobic protective layer prevents water-based developer and fixer solutions from causing lateral erosion of the water-soluble antisolvent layer, thereby ensuring high-precision transfer of the organic pattern.[71] In addition, nanoimprinting is a high-production, cost-effective nanofabrication technique characterized by a streamlined process flow and exceptional capability for large-area, uniform pattern replication. Nanoscale structures can be directly transferred onto photoresist on a substrate using mechanical imprinting with a mold. For example, the transfer of wafer-scale 2D MoS_2 arrays was achieved using a polydimethylsiloxane (PDMS) stamp with a precisely arranged array of microcolumns, and MoS_2-WSe_2 van der Waals heterostructure arrays were subsequently stacked, as shown in Figure 8.6c.[72]

Transparent and flexible electronic devices based on low-dimensional quantum materials can be systematically classified into three categories according to their core functions: transparent and flexible photodetectors, transparent flexible display devices, and transparent flexible energy conversion and storage devices. Transparent and flexible photodetectors are a class of devices that combine optical transparency, mechanical flexibility, and photodetection capabilities. Their primary function is to provide high-fidelity, quantitative detection of optical parameters such as light intensity, wavelength, and polarization state. They play a significant role in practical applications. Leveraging HfO_x/SnO_2 (p–n), SnO_2/ZnO (n–n), and CoO_x/HfO_x (p–p) multi-junction structures, a wearable ultraviolet A (UVA) monitoring device that integrates signal amplification and Bluetooth Low Energy communication capabilities has been successfully developed. As shown in Figure 8.7a, this device can monitor UVA intensity in real time outdoors in sunlight and display the data via a smartphone app, demonstrating its practical potential for personal health protection.[73] In recent years, sustained advances in wearable sensor technology have enabled robust, long-term, in situ monitoring of physiological and biochemical signals, providing critical technological support for personalized medicine. However, existing devices generally employ a single-modality sensing architecture that supports only the collection of a single parameter and the output of isolated data, making

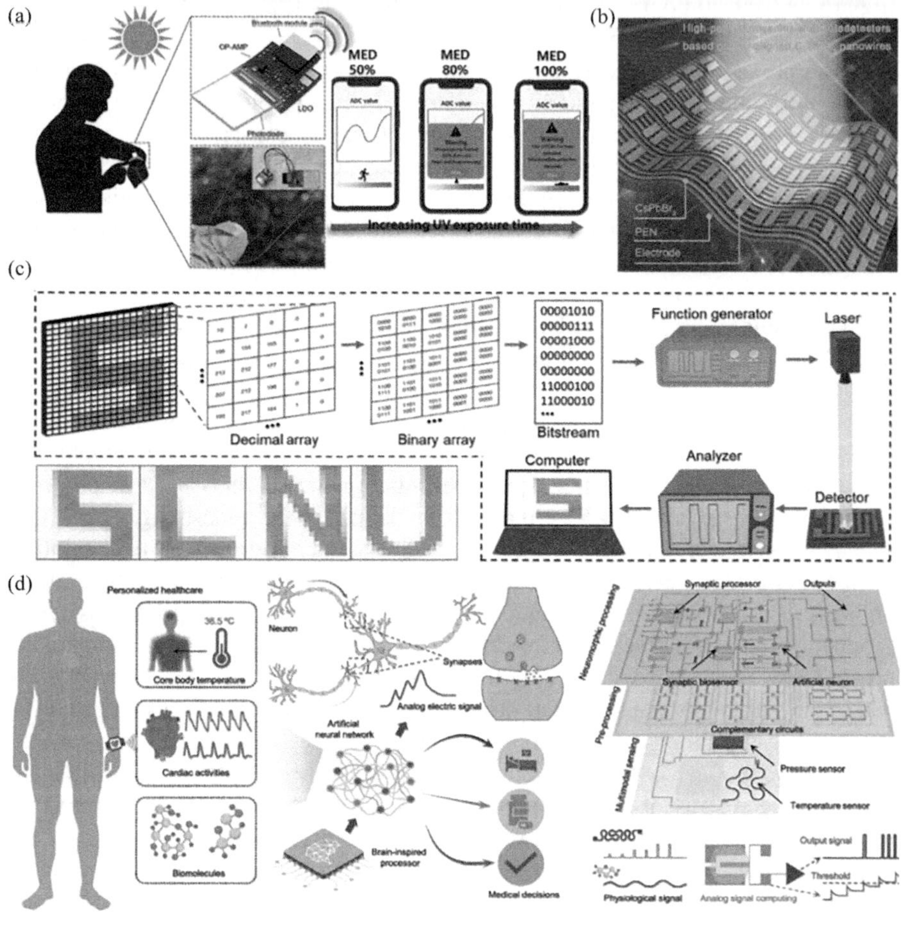

FIGURE 8.7 (a) Wearable UV health monitoring devices for exposure monitoring and warning applications.[73] (b) Schematic illustration of high-performance flexible photodetectors based on self-aligned $CsPbBr_3$ nanowires.[74] (c) Schematic diagram of an optical wireless communication system for image transmission based on $CsPbBr_3$ nanowires flexible photodetectors.[74] (d) Schematic illustrations of the wearable chip-less wearable sensor-processor integrated neuromorphic system.[75]

it difficult to characterize the dynamic coupling relationships between multidimensional physiological parameters (such as the temporal correlation between heart rate variability and cortisol levels). This limits their clinical value in systematic health assessments and early risk warning.

Recently, a research team has developed a wearable chip-less wearable sensor-processor integrated neuromorphic system capable of simultaneously monitoring, processing, and analyzing multimodal physicochemical and biological biomarker data (i.e., metabolites, cardiac activity, and core body temperature), as shown in Figure 8.7d.[75] Furthermore, by directly integrating high-quality $CsPbBr_3$ perovskite nanowires onto a flexible substrate, the resulting seamlessly integrated photodetector exhibits a responsivity of up to 909.9 A W^{-1}, a specific detectivity of 7.35×10^{13} Jones, a fast response time of approximately 50 us, and a wide linear dynamic range of 89.5 dB. Critically, the device maintains robust mechanical stability after withstanding 2,000 bending cycles at a severe bending radius of 0.5 cm, demonstrating its practical capability for high-resolution optical imaging and wireless image transmission, as shown in Figure 8.7b and c.[74]

Transparent flexible display devices encompass transparent flexible organic light-emitting diodes (OLEDs) and transparent flexible QD light-emitting diodes, among others. Their core function is to achieve efficient and uniform electroluminescence in the visible light spectrum while maintaining high optical transmittance and excellent mechanical flexibility. These devices have been deployed in advanced application domains such as foldable displays, augmented reality optical waveguide integration, wearable phototherapy light sources, and dimmable windows for smart buildings. The luminescent performance of these devices is fundamentally constrained by the intrinsic photoluminescence/electroluminescence quantum yields of low-dimensional quantum materials, the equilibrium of carrier injection, as well as the nanoscale topographical uniformity and interfacial stability of multilayer heterostructures on flexible substrates. Transparent flexible OLEDs have emerged as the most technologically mature and commercially viable transparent flexible electroluminescent devices, owing to their self-emissive properties, ultrahigh contrast ratio, microsecond-level response times, and excellent mechanical flexibility. Its core structural innovation lies in replacing traditional ITO electrodes with transparent, flexible conductive electrodes, which, when combined with an organic light-emitting layer and a flexible encapsulation layer, enable flexible lighting and transparent displays. For example, as shown in Figure 8.8a and b, green phosphorescent OLEDs fabricated using flexible transparent conductive electrodes (FTCEs) based on an MXene/AgNWs/MXene sandwich structure have achieved state-of-the-art electroluminescent performance, with a maximum EQE of 22.9% and a current efficiency of 81.4 cd A^{-1}, significantly higher than that of ITO-based reference devices.[76] Furthermore, fully stretchable OLEDs are an indispensable component for skin-mounted display devices. Nevertheless, for over a decade, low power conversion efficiency has remained a critical bottleneck limiting practical applications. Recent research has successfully addressed this challenge by introducing an intrinsic stretchable exciton-exciplex-assisted phosphorescent layer. The fully stretchable OLEDs prepared in this study achieved an EQE of 17.0%, with minimal loss in luminous intensity even under 60% strain, as shown in Figure 8.8c.[77]

Flexible solar cells constitute a core component of transparent and flexible energy conversion and storage devices. Owing to their demonstrated potential in wearable energy system integration and building-integrated photovoltaics, they are widely regarded as one of the most promising next-generation photovoltaic technology pathways.[78–80] In research on high-performance flexible solar cells, designing efficient transparent top electrodes and enhancing light utilization in the active layer are of critical importance. Due to their relatively high surface roughness, traditional ultrathin metal transparent electrodes cannot form good electrical contact with the active layer, resulting in low current and fill factor, which in turn affects device efficiency.

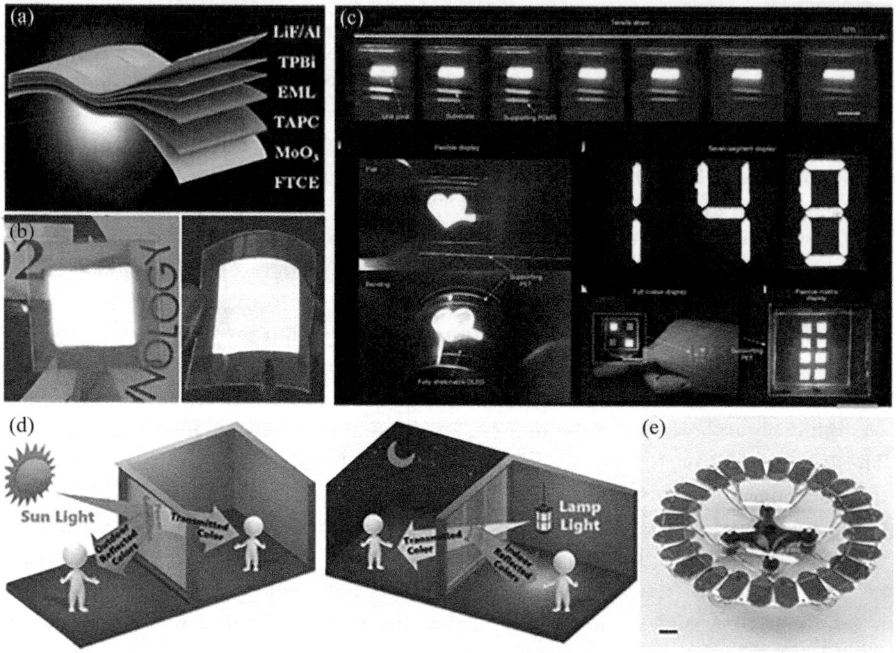

FIGURE 8.8 (a) Diagrams for various FTCEs-based green and red OLEDs.[76] (b) Photographs of PMAM-based green and red flexible OLEDs featuring an emitting area of 25 × 25mm^2.[76] (c) Fully stretchable OLEDs.[77] (d) Diagram of photovoltaic glass window application scenarios during the day and the night.[78] (e) Design of the hybrid-power Solar Hopper quadcopter.[79]

By developing high-performance, ultra-thin, transparent Au/Ag composite top electrodes and incorporating dielectric mirrors, it is possible to improve electrical contact and enhance the re-absorption of light by the active layer, respectively, thereby enhancing cell performance.[81] However, the simultaneous achievement of high power conversion efficiency, exceptional mechanical robustness under extreme deformation conditions, and long-term operational stability remains a fundamental challenge for flexible solar cells. Recent research has reported a strategy that combines data-driven machine learning with passivation methods to utilize the in-situ self-assembly of amorphous β-cyclodextrin derivatives at grain boundaries, thereby simultaneously enhancing the efficiency, mechanical strength, and stability of flexible perovskite solar cells.[82]

8.5 CHALLENGES AND DEVELOPMENT DIRECTIONS

Owing to their unique optical response characteristics, excellent carrier transport properties, high programmability at heterojunctions, and tunable bandgap structures, low-dimensional quantum materials hold great promise for application in next-generation optoelectronic devices designed for low power consumption and ultra-high integration density. The low-power TFETs, high-performance photodetectors, and transparent flexible optoelectronic devices outlined above exhibit compelling application potential in frontier domains, including wearable health-monitoring systems and quantum-information-processing platforms. However, in the transition from laboratory-scale prototypes to industrial applications, there are still bottlenecks in various areas, including material synthesis, device performance optimization, integration processes, and stability.

Although TFETs have made significant progress in breaking through the theoretical limit of 60 mV/dec for SS in traditional MOSFETs and achieving low power consumption, significant challenges remain in terms of performance improvement and industrial application. First, the manufacturing process for TFETs is complex and places high demands on materials and fabrication techniques, which increases both the cost and the difficulty of device production. Second, at the single-device level, fundamental physical trade-offs persist among three cornerstone metrics (i.e., high on-state current, low off-state current, and steep SS), and a unified, manufacturable design framework capable of simultaneously optimizing all three remains elusive. Furthermore, key technologies such as the development of TFET process platforms for circuit-level integration, device matching control, and reliability verification are not yet mature. Therefore, future research should prioritize the development of TFET device architectures enabled by emerging low-dimensional quantum materials (such as topological insulators and staggered magnets), while synergistically integrating multi-physics modulation approaches (strain engineering, gate electric-field control, and exchange-bias magnetic-field tuning) to transcend fundamental performance trade-offs and achieve concurrent optimization of the three defining metrics: on-state current, off-state current, and SS. At the same time, we are exploring new manufacturing processes to reduce process complexity and cost-effective device production.

Although the performance of photodetectors based on low-dimensional quantum materials has been improved by heterostructure fabrication, interface effect modulation, and doping. However, two critical challenges remain in achieving commercial applications: first, the lack of scalable fabrication technologies for large-area, highly uniform photodetector arrays; and second, the absence of effective methods for integrating photodetector arrays with existing silicon electronics technology. Looking ahead, there is an urgent need to overcome the challenges of large-area, uniform fabrication of photodetectors based on low-dimensional quantum materials and their integration with silicon-based processes, in order to advance these devices from laboratory prototypes to large-scale applications. In addition, while most photodetectors currently focus on capturing a single physical quantity (such as light intensity or a specific wavelength), the rapid development of big data analytics, artificial intelligence, and the Internet of Things is driving a shift in application scenarios from "one-dimensional sensing" to "multidimensional integration." Therefore, there is an urgent need to develop high-performance, multidimensional optoelectronic detection technologies to enable the integrated sensing of information such as light intensity, wavelength, and polarization.

Transparent electrodes are one of the critical components of transparent and flexible optoelectronic devices. In recent years, with the rapid development of flexible optoelectronic devices such as wearable devices and curved-screen smartphones, new demands have emerged regarding the mechanical flexibility of transparent electrodes. Traditional transparent electrodes, such as ITO, offer good light transmittance but are highly brittle, making them unsuitable for the repeated bending and stretching required by flexible devices; meanwhile, materials such as metal nanowires and carbon-based materials, while flexible, suffer from low light transmittance or high electrical resistance. The key challenge is achieving low electrical resistance and excellent mechanical flexibility while maintaining high light transmittance. As application demands grow, transparent flexible optoelectronic devices must be capable of multimodal sensing (such as photodetection, temperature sensing, and physiological signal monitoring). Integrating multiple functions into a single device while maintaining high performance and miniaturization poses a significant technical challenge. Looking ahead, it will take synergistic breakthroughs across multiple areas, including materials innovation, structural design, and process optimization, to drive the practical application and industrialization of transparent flexible optoelectronic devices.

REFERENCES

[1] Zhang J, et al. Bottom–up-synthesized graphene nanoribbons for nanoelectronics. *Nature Reviews Materials* 11(3), 194–212 (2026).

[2] Li C, et al. Concurrent pressure-enhanced magnetic order and photoconductivity in quasi-one-dimensional antiferromagnet $MnSb_2S_4$. *Journal of the American Chemical Society* 148(6), 6560–6568 (2026).

[3] Lyu B, et al. Graphene nanoribbons grown in hbn stacks for high-performance electronics. *Nature* 628(8009), 758–764 (2024).

[4] Zhang T, et al. High-quality narrow black phosphorus nanoribbons with nearly atomically smooth edges and well-defined edge orientation. *Nature Materials* 24(9), 1375–1386 (2025).

[5] Cai X, et al. Optical switching of a moiré chern ferromagnet. *Nature* 650(8102), 580–584 (2026).

[6] Dean DJ, et al. Low-power integrated optical amplification through second-harmonic resonance. *Nature* 649(8099), 1159–1164 (2026).

[7] Gao M, et al. Low-dimensional materials for bioelectronic devices. *Nature Reviews Bioengineering* 4(2), 171–191 (2026).

[8] Moore GE. Cramming more components onto integrated circuits. *Proceedings of the IEEE* 86(1), 82–85 (1998).

[9] Frank DJ, et al. Device scaling limits of si mosfets and their application dependencies. *Proceedings of the IEEE* 89(3), 259–288 (2001).

[10] Yang Q, et al. Steep-slope vertical-transport transistors built from sub-5 nm thin van der waals heterostructures. *Nature Communications* 15(1), 1138 (2024).

[11] Ferain I, Colinge CA, Colinge J-P. Multigate transistors as the future of classical metal–oxide–semiconductor field-effect transistors. *Nature* 479(7373), 310–316 (2011).

[12] Yoon JS, Jeong J, Lee S, Baek RH. Bottom oxide bulk finfets without punch-through-stopper for extending toward 5-nm node. *IEEE Access* 7, 75762–75767 (2019).

[13] Shirzaei Sani E, et al. A stretchable wireless wearable bioelectronic system for multiplexed monitoring and combination treatment of infected chronic wounds. *Science Advances* 9(12), eadf7388 (2023).

[14] Shin H-S, et al. Soft, Skin-interfaced wireless electrogoniometry systems for continuous monitoring of finger and wrist joints. *Nature Communications* 16(1), 4426 (2025).
[15] Hu S, et al. Transparent artificial intelligence-enabled interpretable and interactive sleep apnea assessment across flexible monitoring scenarios. *Nature Communications* 16(1), 7548 (2025).
[16] Shi X, et al. Large-area display textiles integrated with functional systems. *Nature* 591(7849), 240–245 (2021).
[17] Zhu Z, et al. An immunocompatible conductive hydrogel via anion-π interlocking as an injectable bridge for sustained bioelectronic interfacing. *Advanced Materials*, 38(21), e72820 (2026).
[18] Zhou J, et al. A bioresorbable and bioimplantable energy harvesting-storage integrated system as wireless power supply for biomedical electronics. *Advanced Materials* 38(12), e20861 (2026).
[19] Yin L, et al. Two-dimensional semiconductors and transistors for future integrated circuits. *ACS Nano* 18(11), 7739–7768 (2024).
[20] Chhowalla M, Jena D, Zhang H. Two-dimensional semiconductors for transistors. *Nature Reviews Materials* 1(11), 16052 (2016).
[21] Choi WY, Park BG, Lee JD, Liu TJK. Tunneling field-effect transistors (TFETs) with subthreshold swing (SS) less than 60 mv/dec. *IEEE Electron Device Letters* 28(8), 743–745 (2007).
[22] Sarkar D, et al. A subthermionic tunnel field-effect transistor with an atomically thin channel. *Nature* 526(7571), 91–95 (2015).
[23] Ionescu AM, Riel H. Tunnel field-effect transistors as energy-efficient electronic switches. *Nature* 479(7373), 329–337 (2011).
[24] Turkane S, Kureshi AK. Review of tunneling field-effect transistors (TFETs). *International Journal of Applied Engineering Research* 11, 4922–4929 (2016).
[25] Esseni D, et al. A review of selected topics in physics based modeling for tunnel field-effect transistors. *Semiconductor Science and Technology* 32(8), 083005 (2017).
[26] Luisier M, Klimeck G. Simulation of nanowire tunneling transistors: From the wentzel–kramers–brillouin approximation to full-band phonon-assisted tunneling. *Journal of Applied Physics* 107(8), 084507 (2010).
[27] Cao W, et al. Subthreshold-swing physics of tunnel field-effect transistors. *AIP Advances* 4(6), 067141 (2014).
[28] Weber WM, Mikolajick T. Silicon and germanium nanowire electronics: Physics of conventional and unconventional transistors. *Reports on Progress in Physics* 80(6), 066502 (2017).
[29] Balandin AA. Phononics of graphene and related materials. *ACS Nano* 14(5), 5170–5178 (2020).
[30] Razaq A, et al. Review on graphene-, graphene oxide-, reduced graphene oxide-based flexible composites: From fabrication to applications. *Materials* 15(3), 1012 (2022).
[31] Tyulnev I, et al. Valleytronics in bulk MoS_2 with a topologic optical field. *Nature* 628(8009), 746–751 (2024).
[32] Ovchinnikov D, et al. Electrical transport properties of single-layer WS_2. *ACS Nano* 8(8), 8174–8181 (2014).
[33] Hou J, et al. Treating acute kidney injury with antioxidative black phosphorus nanosheets. *Nano Letters* 20(2), 1447–1454 (2020).
[34] Zong X, et al. Black phosphorus-based van der waals heterostructures for mid-infrared light-emission applications. *Light: Science & Applications* 9(1), 114 (2020).
[35] Chen X, et al. 2D steep-slope tunnel field-effect transistors tuned by van der waals ferroelectrics. *Advanced Electronic Materials* 11(4), 2400463 (2025).

[36] Li L, et al. 2D short-channel tunneling transistor relying on dual-gate modulation for integrated circuits application. *Advanced Functional Materials* 33(45), 2304591 (2023).
[37] Xiong X, et al. A transverse tunnelling field-effect transistor made from a van der waals heterostructure. *Nature Electronics* 3(2), 106–112 (2020).
[38] Kim S, et al. Thickness-controlled black phosphorus tunnel field-effect transistor for low-power switches. *Nature Nanotechnology* 15(3), 203–206 (2020).
[39] Wang M, et al. Silicon-based intermediate-band infrared photodetector realized by te hyperdoping. *Advanced Optical Materials* 9(4), 2001546 (2021).
[40] Casalino M, et al. Near-infrared all-silicon photodetectors. *International Journal of Photoenergy* 2012(1), 139278 (2012).
[41] Guo Q, et al. Black phosphorus mid-infrared photodetectors with high gain. *Nano Letters* 16(7), 4648–4655 (2016).
[42] Xie Y, et al. Ultrabroadband MoS_2 photodetector with spectral response from 445 to 2717 nm. *Advanced Materials* 29(17), 1605972 (2017).
[43] Zhang J, et al. Ultrabroadband detection and self-powered functionality in quasi-one-dimensional $Nb_3Se_{12}I$ nanowire photodetectors for bionic vision applications. *ACS Applied Materials & Interfaces* 17(14), 21448–21458 (2025).
[44] Kim B, et al. Ultrahigh-gain colloidal quantum dot infrared avalanche photodetectors. *Nature Nanotechnology* 20(2), 237–245 (2025).
[45] Feng M, et al. Overcoming size limits with dynamic templates enabling large area single crystal nanowire arrays for photodetectors. *Nature Communications* 16(1), 9891 (2025).
[46] Mu Y, et al. Homo-type α-In_2Se_3/$PdSe_2$ ferroelectric van der waals heterojunction photodetectors with high-performance and broadband. *Advanced Functional Materials* 34(27), 2315543 (2024).
[47] Wu L, et al. Polarity-switchable and self-driven photo-response based on vertically stacked type-iii GeSe/SnS_2 heterojunction. *Advanced Materials Interfaces* 9(12), 2102099 (2022).
[48] Gao M, et al. Interface engineering of van der Waals devices based on metal disulfide vertical heterostructures for enhanced photoresponse. *Nano Letters* 26(1), 134–141 (2026).
[49] Jeong S-J, et al. High-gain Ag_2Te/MoS_2 hybrid photodetectors for short-wave infrared imaging. *Advanced Materials*, 38(22), e20984 (2026).
[50] Dat VK, et al. Van der waals integration of 1D $Nb_2Pd_3Se_8$ and 2D WSe_2 for gate-tunable in-sensor image processing. *Advanced Materials* 37(43), e00011 (2025).
[51] Li Y, et al. Ultrafast 0D/1D ZnO/CuO photodetector in nanosecond scale by engineering the type-ii heterostructure. *Applied Physics Letters* 126(8), 083503 (2025).
[52] Han J, et al. Carbon nanotube transistor with colloidal quantum dot photosensitive gate for ultrahigh external quantum efficiency photodetector. *ACS Nano* 17(10), 9510–9520 (2023).
[53] Li L, et al. Zno quantum dot decorated Zn_2SnO_4 nanowire heterojunction photodetectors with drastic performance enhancement and flexible ultraviolet image sensors. *ACS Nano* 11(4), 4067–4076 (2017).
[54] Zhang T, et al. High performance metal-semiconductor-metal ultraviolet photodetector based on mixed-dimensional TiO_2/$CsPbBr_3$ heterostructures. *Physica Scripta* 99(1), 015526 (2024).
[55] Ahmad W, et al. Interface engineering with graphene contacts in VP/MoS_2 heterostructures for high-performance, durable, and multispectral photodetection. *ACS Applied Materials & Interfaces* 18(11), 16767–16777 (2026).
[56] Li J, et al. Ultrafast self-driven WSe_2 photodetectors with bottom Schottky contacts. *Advanced Science* 12(40), e10373 (2025).

[57] Mao J, et al. Conformal MoS_2/Silicon nanowire array heterojunction with enhanced light trapping and effective interface passivation for ultraweak infrared light detection. *Advanced Functional Materials* 32(11), 2108174 (2022).
[58] Zhu Q, et al. High-performance near-infrared $PtSe_2$/n-Ge heterojunction photodetector with ultrathin Al_2O_3 passivation interlayer. *Science China Materials* 66(7), 2777–2787 (2023).
[59] Ye P, et al. Si-cmos-compatible 2D $PtSe_2$-based self-driven photodetector with ultrahigh responsivity and specific detectivity. *Science China Materials* 66(1), 193–201 (2023).
[60] Lu C, et al. Direct z-scheme $SnSe_2$/SnSe heterostructure passivated by Al_2O_3 for highly stable and sensitive photoelectrochemical photodetectors. *ACS Applied Materials & Interfaces* 15(4), 6156–6168 (2023).
[61] Yue Z, et al. Ultrafast and self-driven flexible photodetector based on vertical MoS_2/Si heterojunction through enhanced light-trapping structures and Al_2O_3 interface passivation. *Applied Surface Science* 655, 159630 (2024).
[62] Song C, et al. Ultra-broadband polarization-sensitive O-doped $ZrTe_3$ photodetector induced by uv laser. *Advanced Materials* 37(40), 2506840 (2025).
[63] Wang L, et al. Ultrastable colloidal quantum dots-based photoelectrochemical photodetectors for weak-light underwater optical communication. *Nano Research* 18(11), 94907851 (2025).
[64] Su J, et al. Flexible and broadband photodetector based on mn-doped ZnSe on paper substrate. *Optical Materials* 168, 117443 (2025).
[65] Hu Y, et al. A controllable doping method for Ga_2O_3 film construction and optimized density for high-performance photodetectors. *Applied Physics Letters* 126(6), 063303 (2025).
[66] Fukuda K, et al. Fully-printed high-performance organic thin-film transistors and circuitry on one-micron-thick polymer films. *Nature Communications* 5(1), 4147 (2014).
[67] Wang D, et al. Large area polymer semiconductor sub-microwire arrays by coaxial focused electrohydrodynamic jet printing for high-performance OFETs. *Nature Communications* 13(1), 6214 (2022).
[68] Sun L, et al. All-solution-processed ultraflexible wearable sensor enabled with universal trilayer structure for organic optoelectronic devices. *Science Advances* 10(15), eadk9460 (2024).
[69] Tan P, et al. Solution-processable, soft, self-adhesive, and conductive polymer composites for soft electronics. *Nature Communications* 13(1), 358 (2022).
[70] Bae K, et al. Washable heat-resistant and inkjet-printed devices on cotton fabric for wearable applications. *Nature Communications* 16(1), 8615 (2025).
[71] Ni Y, et al. Superintegrated conformable organic transistors based on a universal microlithographic strategy. *Nature Communications* 16(1), 9246 (2025).
[72] Liu L, et al. A mass transfer technology for high-density two-dimensional device integration. *Nature Electronics* 8(2), 135–146 (2025).
[73] Kim YB, et al. Transparent UVA photodetectors based on oxide semiconductors for real-time wearable skin protection monitoring. *Science Advances* 11(42), eaea7218 (2025).
[74] Mao W, et al. Nanogroove-guided $CsPbBr_3$ nanowire arrays on bendable plastic enable high-performance flexible photodetectors for image sensing and transmission. *ACS Applied Materials & Interfaces* 18(9), 14157–14168 (2026).
[75] Choi Y, et al. All-printed chip-less wearable neuromorphic system for multimodal physicochemical health monitoring, *Nature Communications* 16(1), 5689 (2025).
[76] Yang Z, et al. Mxene agnws mxene sandwich-structured transparent electrode for high-performance flexible OLEDs. *Small* 21(14), 2409621 (2025).

[77] Zhou H, et al. Exciplex-enabled high-efficiency, fully stretchable OLEDs. *Nature* 649(8097), 604–611 (2026).
[78] Tian Y, et al. Organic solar windows with full visual aesthetics. *Science China Materials* 69(4), 1986–1994 (2025).
[79] Hailegnaw B, et al. Flexible quasi-2D perovskite solar cells with high specific power and improved stability for energy-autonomous drones. *Nature Energy* 9(6), 677–690 (2024).
[80] Wang T, et al. Scalable semitransparent organic solar cells with robust film thickness tolerance for building-integrated photovoltaics. *Nature Communications* 17(1), 2916 (2026).
[81] Xu G, et al. High-performance colorful semitransparent polymer solar cells with ultra-thin hybrid-metal electrodes and fine-tuned dielectric mirrors. *Advanced Functional Materials* 27(15), 1605908 (2017).
[82] He M, et al. Amorphous grain boundary engineering for scalable flexible perovskite photovoltaics with improved stability. *Nature Energy* 11(2), 262–273 (2026).

9 Applications of Low-Dimensional Quantum Materials in Energy Conversion and Catalysis

Yundan Liu, Lamei Tan, Haoyan Zhou, and Zongyu Huang

9.1 BACKGROUND AND MOTIVATION

Driven by the continuous increase in global energy demand and the urgent need to reduce carbon emissions, the development of sustainable energy conversion technologies has become a key research focus. In this context, efficient utilization of renewable energy and conversion of small molecules such as water and carbon dioxide are considered promising routes toward a low-carbon energy system. Achieving these processes relies critically on the design and optimization of highly active and stable catalytic materials, which has stimulated extensive research efforts in recent years. However, the challenge lies in the fact that existing catalytic materials struggle to strike a balance between activity, selectivity, stability, and cost. Although precious metals (such as Pt, Ir, and Ru) exhibit excellent performance, their scarcity and high cost severely limit their large-scale application. Therefore, the development of highly efficient, stable, and abundant non-precious metal catalytic materials has become a focal point of research at the intersection of materials science and catalytic chemistry.[1,2]

Traditional bulk catalytic materials (such as metal oxides, sulfides, and carbon-based materials) have accumulated a wealth of knowledge through long-term research; however, their inherent structural characteristics limit further improvements in performance. First, regarding active site utilization, catalytic reactions in bulk materials primarily occur on the outer surface, leaving a large number of internal atoms unable to participate in the reaction. This results in low active site density and poor atomic economy.[3] Second, during photocatalysis, photogenerated electron-hole pairs are prone to bulk recombination within the material, and the long-distance carriers must travel to reach surface reaction sites, making it difficult to improve quantum efficiency.[4] Furthermore, the band structure and surface chemistry of bulk materials are difficult to flexibly control after synthesis, limiting the scope for optimizing catalytic performance through methods such as interface engineering and

DOI: 10.1201/9781003655305-12

defect engineering. These limitations have prompted researchers to turn their attention to low-dimensional quantum materials with unique physicochemical properties.

9.1.1 Types and Characteristics of Low-Dimensional Quantum Materials

Low-dimensional quantum materials are functional materials in which electronic states and physical properties are significantly influenced by quantum effects. They possess highly unique electronic characteristics and exhibit pronounced quantum confinement effects due to their reduced dimensionality. Based on their spatial dimensions, they can be classified into zero-dimensional (0D), one-dimensional (1D), and two-dimensional (2D) materials.[5]

In 0D quantum materials, electrons are confined in all three dimensions (3D) and cannot move freely. Due to quantum confinement effects, their electronic density of states (DOS) exhibits a discrete energy level structure. Typical zero-dimensional materials include carbon quantum dots, graphene quantum dots, and transition metal sulfide quantum dots. Carbon quantum dots typically consist of a core formed by sp^2/sp^3 hybridized carbon atoms and an amorphous shell rich in O/N functional groups, and they have widespread applications in the field of catalysis. 1D quantum materials include nanowires, nanorods, and nanotubes, in which electrons are confined in 2D. Typical examples include carbon nanotubes and metal oxide nanowires. 1D materials exhibit excellent charge transport properties along the axial direction, while the radial quantum confinement effect allows for the regulation of their electronic structure. 2D quantum materials, represented by nanosheets and thin films, confine electrons in only 1D. Since the discovery of graphene in 2004, the family of 2D materials has expanded rapidly to include transition metal dichalcogenides (TMDs), MXenes, black phosphorus, and layered double hydroxides.[6] Among these, graphene, as a representative of 2D materials, possesses extremely high carrier mobility, excellent mechanical strength, and good electrical conductivity. TMDs (such as molybdenum disulfide [MoS_2] and WS_2), in contrast, have garnered significant attention in the field of catalysis due to their tunable bandgaps and abundant edge active sites.

The key quantum properties of these materials are briefly summarized as follows:

1. High Surface Area and Edge Effects. The atomic-scale thickness of 2D materials means that nearly every atom is exposed at the surface. In particular, edge sites often exhibit higher intrinsic reactivity than the basal plane due to their undercoordinated state.[7]
2. Tunable Band Structure. By varying the number of layers or applying an external field, phase transitions from metallic to semiconducting states and even to superconductive states can be achieved. For example, a single layer of MoS_2 has a direct bandgap of 1.67 eV, whereas the bulk material exhibits an indirect bandgap.
3. Strong Quantum Coupling. In these low-dimensional systems, interactions between energy levels, spin, excitons, and phonons are significantly enhanced. This tunability of multi-field coupling enables materials to

flexibly adjust the kinetics of their excited-state processes to meet specific photocatalytic or electrocatalytic requirements.

4. Strong Excitonic Effects. In low-dimensional quantum materials (particularly 2D TMDs), due to the significant weakening of the dielectric screening effect and strong quantum confinement, electrons and holes combine via Coulombic interactions to form tightly bound "excitons." Unlike the free excitons with extremely low energy in conventional bulk semiconductors, the binding energy of excitons in low-dimensional materials can reach hundreds of meV. These strong excitonic effects not only dominate the material's light absorption process but also determine the funnel-like transfer of excited-state energy and the dynamics of charge separation at interfaces.[8]
5. Topologically Nontrivial States. Some low-dimensional quantum materials possess topological surface or edge states protected by symmetry. These electronic states exhibit high carrier mobility without backscattering and demonstrate exceptional robustness to localized lattice defects and impurities, providing a revolutionary foundation for maintaining efficient, stable charge transport in complex electrochemical environments.
6. Valleytronics and Spin–Valley Locking. In addition to the traditional charge and spin degrees of freedom, 2D quantum materials (such as monolayer WSe_2 and WS_2) exhibit a unique "valley" degree of freedom due to the breaking of inversion symmetry and strong spin–orbit coupling (SOC) (i.e., local extrema at the bottom of the conduction band or the top of the valence band in momentum space). This spin–valley locking property allows us to selectively excite carriers in specific energy valleys using circularly polarized light, providing a completely new space for the physical parameter control of asymmetric catalytic reactions with extremely high spin/valley selectivity.

9.1.2 Unique Advantages for Energy and Catalytic Applications

Low-dimensional quantum materials offer precise control over microscopic electronic states, demonstrating advantages in the fields of energy and catalysis that traditional bulk materials cannot match. First, the highly engineered electronic structure allows the Fermi level, work function, and bandgap to be optimized according to the adsorption energy of reaction intermediates. For 2D transition metal sulfides, altering the number of layers enables a transition from an indirect bandgap to a direct bandgap. Second, the exceptional carrier dynamics of low-dimensional materials are crucial for optoelectronic energy conversion. In photocatalysis, the ultra-thin nature of low-dimensional materials can significantly shorten the distance that photogenerated carriers must travel from the bulk to the surface, thereby substantially suppressing bulk recombination. This facilitates electron-hole separation and interfacial injection, ultimately enhancing energy conversion efficiency.[4] In traditional composite photocatalysts, highly conductive materials such as graphene serve merely as electron transport media and do not possess catalytically active sites of their own. In contrast, topological Weyl semimetals (such as transition metal phosphides) can utilize their bulk Dirac bands to provide extremely high carrier mobility at room temperature, while their topologically protected surface electronic states directly function as highly active catalytic sites. This intrinsic integration of "ultra-high bulk electron transport" and "high intrinsic surface activity" effectively eliminates charge transfer barriers at the composite interfaces, providing a novel physical approach to addressing the issue of charge recombination. Finally, the ease of constructing heterojunctions offers unlimited possibilities for the design of multifunctional catalytic systems. Since the surfaces of two-dimensional materials lack dangling bonds, different layered materials can be stacked via van der Waals forces to form van der Waals heterojunctions that are not constrained by lattice matching. Such heterojunctions can generate strong built-in electric fields that drive the directional migration of charges.[9]

9.1.3 Scope of This Chapter

Against this backdrop, this chapter aims to provide an overview of research progress in low-dimensional quantum materials for energy conversion and catalysis. Section 9.2 focuses on energy conversion applications, including photocatalysis (water splitting and CO_2 reduction), electrocatalysis (hydrogen evolution reaction [HER], oxygen evolution reaction [OER], oxygen reduction reaction [ORR], electrochemical carbon dioxide reduction reaction [CO_2RR]), and structure–property relationships; Section 9.3 delves into catalytic mechanisms; and Section 9.4 summarizes current challenges and outlines future research directions.

9.2 ENERGY CONVERSION

9.2.1 Photocatalytic Energy Conversion

Photocatalytic technology harnesses solar energy to drive chemical reactions, making it a key approach to achieving sustainable energy conversion. Due to their unique optoelectronic properties, low-dimensional quantum materials have demonstrated tremendous potential in fields such as photocatalytic water splitting for hydrogen production and CO_2 photoreduction.

Since 1967, when Fujishima, then a first-year master's student, discovered under Honda's guidance that TiO_2 electrodes could decompose water under ultraviolet light, many researchers have conducted extensive studies on water splitting using semiconductor photoelectrodes and photocatalysts. Research on photocatalytic water splitting for hydrogen production has evolved over several decades. The fundamental process of photocatalytic water splitting involves three key steps: light absorption, separation and migration of photogenerated charge carriers, and surface redox reactions.[10] In low-dimensional materials, this process exhibits significant quantum dynamic characteristics, as shown in Figure 9.1.

1. Light Absorption. When a semiconductor material absorbs a photon with energy greater than its bandgap (E_g), an electron-hole pair is generated. However, in 2D materials, the generated carriers are not fully free electrons and holes, but rather strongly bound excitons. The extremely high DOSs

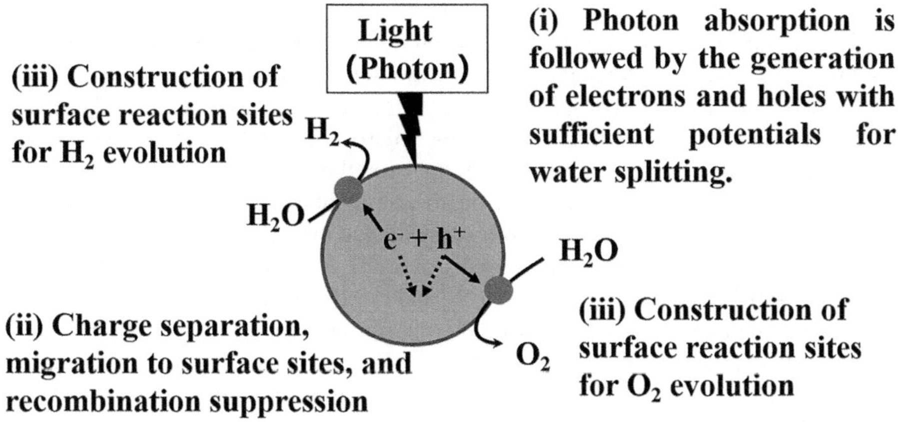

FIGURE 9.1 Main processes in photocatalytic water splitting.

and strong light-matter interactions endow 2D materials with exceptional light-harvesting capabilities.

2. Separation and Migration of Photogenerated Carriers. In low-dimensional materials, the separation of photogenerated carriers and their migration toward surface reaction sites is typically facilitated by internal polarization electric fields, defect trapping, or band offsets at heterojunctions. In this process, the ultra-thin nature of 2D materials reduces the migration distance of charges to the surface to the atomic scale, greatly suppressing bulk recombination.
3. Surface Redox Reactions. Dissociated electrons reduce water to produce H_2, while holes oxidize water to produce O_2. The abundant surface states and tunable work function of low-dimensional materials enable efficient injection of charge into the molecular orbitals of adsorbed reactants.

Photocatalytic water splitting for hydrogen production can be carried out using either suspended particles or flat-plate configurations, with the latter being simpler and more convenient. Professor Domen's team in Japan successfully eliminated charge recombination losses using a flat-plate reactor based on Al-doped $SrTiO_3$. They achieved an external quantum efficiency as high as 96% at wavelengths of 350–360 nm, demonstrating that powder catalysts can also achieve near-perfect conversion efficiency.[11]

9.2.2 Electrocatalytic Energy Conversion

Electrocatalysis plays a central role in clean energy conversion, involving key processes such as the HER, the OER, the ORR, and the CO_2RR. Due to their high electrical conductivity, abundance of active sites, and structural tunability, low-dimensional quantum materials have emerged as promising candidates for replacing precious metal catalysts.

Pt-group metals are currently the most effective HER catalysts; however, their high cost and scarcity have driven the development of non-precious metal catalysts. Notably, 2D TMDs (particularly MoS_2) are among the most extensively studied non-precious metal HER catalysts. Jaramillo et al. were the first to conclusively demonstrate, by combining scanning tunneling microscopy (STM) with macroscopic electrochemical testing, that the active sites for the HER on nanoscale MoS_2 are located at its edge sites rather than on the basal plane. Studies have shown that regardless of whether the samples were annealed at 400°C or 550°C, and regardless of the size of the nanoparticles, the reaction current density exhibits a strict linear relationship with the edge length of MoS_2, and is independent of its total surface area coverage. The experimentally measured edge site activity is in excellent agreement with the hydrogen adsorption free energy (ΔG_H*) calculated by density functional theory (DFT), confirming that MoS_2 follows the classic catalytic "volcano plot" pattern, with activity second only to platinum-group noble metals. This discovery not only experimentally unravels the catalytic sites of MoS_2 but also lays the theoretical and experimental foundation for designing highly efficient, low-cost non-precious metal HER catalysts by maximizing edge site exposure or regulating the edge electronic structure.

In addition to traditional transition metal chalcogenides that rely on edge sites or defect engineering, topological quantum materials have recently pioneered an entirely new "topological catalysis" paradigm in the field of electrocatalysis. In traditional catalytic materials, active surface sites are highly susceptible to deactivation due to oxidation, restructuring, or poisoning when exposed to strong acid/base electrolytes or during long-term operation. In contrast, the surface states (topological surface states [TSS]) of topological materials are protected by the symmetry of the topological invariants of the host band structure. Taking 2D topological materials (such as few-layer $ZrTe_5$ or 1T′-$MoTe_2$) as examples, their protected surface or edge states can provide nearly lossless electronic transport pathways. When participating in HER or OER reactions, these spin-polarized topological electronic states not only significantly reduce the contact resistance for charge transfer from the electrode to reaction intermediates but also optimize the adsorption/desorption kinetics of intermediates (such as H*) through the spin-momentum locking effect. Due to the "immunity" (robustness) of topological states, the main pathways for catalytic activity remain unobstructed even in the presence of vacancies, impurities, or localized amorphization on the material surface. This mechanism, which directly couples microscopic quantum topological invariants to macroscopic electrochemical reaction kinetics, offers a unique pathway for developing energy conversion materials with extreme lifetimes.

A recent high-throughput computational study based on DFT identified 11 high-potential topological catalysts, as shown in Figure 9.2. Notably, Al_1Pd_1(001), Co_1Ge_1(001), Ni_1Si_1(001), and Ga_2Sc_1(110), have Gibbs free energies for hydrogen evolution are extremely close to the ideal values (−0.16, −0.16, −0.18, and −0.17 eV, respectively). The electrocatalytic activity of topological materials is dominated by their unique macroscopic quantum states, including non-trivial band topologies, symmetry-protected gapless surface states, and surface Fermi arcs. These TSSs, which are not easily disrupted by microscopic local defects, can directly participate

(a)

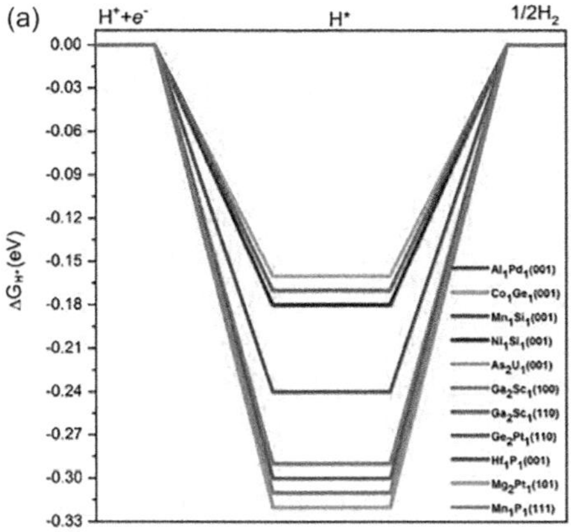

(b)

Surface	ΔG_H (eV)
Al_1Pd_1(001)	-0.16
Co_1Ge_1(001)	-0.16
Mn_1Si_1(001)	-0.24
Ni_1Si_1(001)	-0.18
As_2U_1(001)	-0.29
Ga_2Sc_1(100)	-0.29
Ga_2Sc_1(110)	-0.17
Ge_2Pt_1(110)	-0.30
Hf_1P_1(001)	-0.31
Mg_2Pt_1(101)	-0.32
Mn_1P_1(111)	-0.31

FIGURE 9.2 (a) DFT calculated free energy diagrams for HER on the 11 promising topological quantum materials with their corresponding surface termination. Some curves overlap due to similar ΔG_H* values among different materials and surface terminations. (b) Summary table listing the corresponding adsorption surfaces and their calculated ΔG_H* values. A ΔG_H* value close to 0 eV is regarded as optimal for effective HER catalytic activity. HER, Hydrogen evolution reaction; DFT, density functional theory; ΔG_H*, Gibbs free energy of hydrogen adsorption.

in reaction kinetics as high-speed electron transport channels, thereby creating a broad and excellent range of hydrogen adsorption free energies (e.g., the ΔG_H* of Al_1Pd_1(001) and Co_1Ge_1(001) reach −0.16 eV).[12]

9.2.3 Structure–Property Relationships in Energy Conversion

In energy conversion systems, the macroscopic electrochemical and optoelectronic properties of low-dimensional materials are deeply dependent on the precise arrangement of their microscopic atomic structures and electronic states. Through multidimensional structural engineering, it is possible to effectively overcome the inherent physicochemical limitations of these materials.

9.2.3.1 Layer-Number Effect

When bulk materials are continuously exfoliated and thinned along a specific direction until they approach the 2D limit of a single atomic or molecular layer, their physicochemical properties undergo a dramatic transformation. This phenomenon is known as the "Layer-Number Effect." The root cause of this effect lies in the drastic weakening or even disappearance of interlayer van der Waals interactions along the z-axis, as well as the resulting strong quantum confinement effect.

From the perspective of electronic structure, the reduction in dimensionality directly alters the periodic boundary conditions of the crystal lattice, leading to a fundamental restructuring of the DOS distribution within the Brillouin zone. This restructuring not only causes significant changes in the bandgap width but can even trigger a transition from an indirect bandgap to a direct bandgap. Geometrically, a reduction in the number of layers implies an exponential increase in the material's specific surface area, exposing atoms previously hidden within the bulk as active surface sites. In the field of energy conversion and storage, the dual advantages of "electronics" and "geometry" brought about by the number of layers are key to enhancing photocatalytic efficiency and ion transport kinetics.

The modulation of electronic band structures by the number of layers is vividly demonstrated in the photocatalytic and photovoltaic applications of TMDs. Taking the classic example of MoS_2, the pioneering work by Mak et al. clearly elucidated this process. Thermodynamically stable bulk MoS_2 is an indirect-bandgap semiconductor, with its conduction band bottom located midway between the Γ point and the K point in the Brillouin zone, while the valence band top lies at the Γ point, and the bandgap width is approximately 1.2 eV. This indirect bandgap characteristic requires electrons to rely on phonons to satisfy momentum conservation during transitions, resulting in extremely low light absorption and luminescence efficiencies, which severely limits its application in photocatalysis. However, when MoS_2 is exfoliated into a single layer, the disappearance of interlayer interactions causes a dramatic shift in the energy of the orbitals near the Γ point (primarily composed of the Mo d_{z^2} and S p_z orbitals). Ultimately, both the bottom of the conduction band and the top of the valence band shift to the K-point, transforming monolayer MoS_2 into a direct-bandgap semiconductor with a bandgap of approximately 1.8 eV. This fundamental restructuring of the band structure greatly enhances the interaction between light and matter, enabling the monolayer material to exhibit extremely high light absorption and favor its applications in high-efficiency solar energy conversion and photocatalytic water splitting.[42]

The number-of-layers effect not only alters the band structure but can also induce topological phase transitions. Taking the transition-metal dichalcogenide $MoTe_2$ as an example, when its thickness is reduced from the bulk phase to a single layer or a few layers, the material exhibits non-trivial Z_2 topological invariance. This topological quantum state transition, induced by the reduction in dimensionality, enables the few-layer 1T′-$MoTe_2$ to exhibit significantly higher activity in the HER compared to its topologically trivial phase (2H phase).

9.2.3.2 Defect Effects

Point defects (such as vacancies and interstitial atoms) or line defects (such as grain boundaries) in a crystal lattice disrupt the local periodic potential, introducing new mid-bandgap states.[43] In traditional solid-state physics, lattice defects are often viewed as centers of electron scattering or traps for nonradiative recombination, exerting a negative influence on the macroscopic properties of materials (such as carrier mobility). However, in energy conversion applications, "perfect" does not always equate to "highly efficient." For many low-dimensional quantum materials (such as 2D transition metal chalcogenides or carbon-based nanomaterials), the highly symmetric and thermodynamically stable basal plane typically exhibits chemical inertness. This is because the perfect lattice structure lacks atoms with unsaturated coordination, resulting in ΔG_H^* for adsorption of reaction intermediates that is either too high or too low, making it difficult to satisfy the Sabatier principle.

The core of defect engineering lies in introducing point defects (such as vacancies or interstitial atoms) or line defects into the crystal lattice through controlled methods (such as plasma etching, thermal annealing, or chemical stripping). From the perspective of electronic structure evolution, the introduction of defects disrupts the local periodic potential, thereby generating new mid-gap states within the originally well-defined bandgap. These defect energy levels not only act as shallow potential wells for electrons or holes, enhancing the material's overall electrical conductivity, but also fundamentally rearrange the charge distribution and spin states of local atoms, thereby awakening the originally inert basal plane and creating high-density catalytic or energy-storage active sites. The pioneering work by Xie et al. vividly demonstrates the power of defect engineering. By controlling the addition of excess thiourea, Xie et al. successfully prepared defect-rich ultrathin MoS_2 nanosheets. The highly dense lattice defects disrupt the originally long-range ordered basal plane, causing microscopic cracking of the basal plane and thereby exposing a large number of additional active edge sites on the originally catalytically inert 2D basal plane. This structural reorganization significantly increases the density of catalytically active sites, ultimately resulting in an ultra-low overpotential of only 120 mV and a massive exchange current density in macroscopic electrochemical testing.[13]

Furthermore, when defect engineering intersects with the topological quantum states of materials, it sparks a higher-order revolution in energy conversion technology. Beyond applications in spintronics and quantum information processing, the most attention-grabbing and industrially promising "breakout" research area in topological quantum materials today is undoubtedly topological catalysis. Traditional multi-phase electrocatalytic technologies (such as HER and OER in water electrolysis) rely heavily on the d-band centers of noble metals (such as Pt, Ir, and Ru) for regulation. This classical paradigm currently faces serious bottlenecks, including limited catalyst active site density, high resistance to interfacial electron transfer, weak resistance to surface environmental contamination, and poor long-term stability under high current densities.

Quantum materials, represented by topological semimetals (including Dirac semimetals, Weyl semimetals, and nodal semimetals), offer a novel physical approach to this century-old chemical challenge. The intrinsic electrocatalytic advantages of topological semimetals stem from the perfect alignment between their topological features in momentum space and the kinetics of surface electrochemical reactions. Unlike traditional semiconductors or insulators, topological semimetals possess a linearly crossed band structure in the bulk. This electronic dispersion relation, analogous to the relativistic Dirac equation, endows the material with an extremely low effective electron mass and ultrahigh electronic mobility, fundamentally eliminating the internal charge transfer resistance from the external circuit electrodes to the catalytic surface. Topological Fermi arcs exist in Weyl systems with time-reversal or inversion- symmetry breaking, while in 3D nodal semimetals (such as the TiSi system and $Co_3Sn_2S_2$), polarized flat bands exist as shown in Figure 9.3.[14]

These surface states not only provide an exceptionally high carrier density (with a strong DOS peak) but also offer abundant active sites for molecular adsorption. More crucially, these non-trivial surface states are strictly protected by global topological

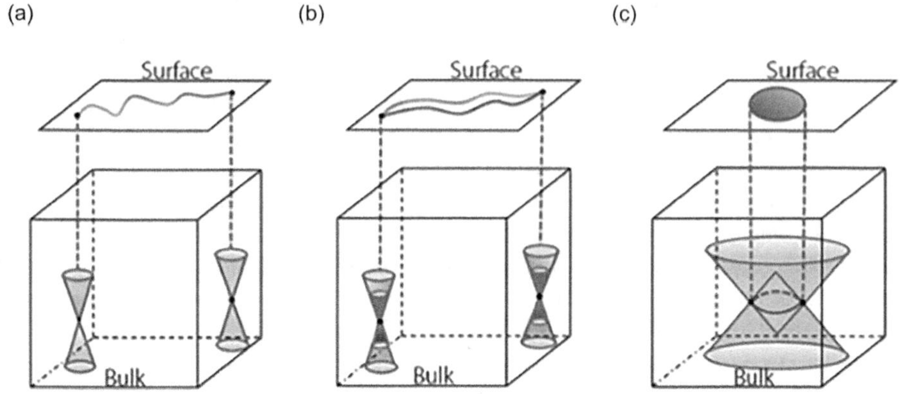

FIGURE 9.3 (a) The schematic diagram of a Weyl semimetal with one pair of Weyl points in the bulk and a Fermi arc on the surface. The schematic diagram shown in (b) is similar to that of (a) but represents a Dirac semimetal. Weyl points have double band degeneracy, while Dirac points have fourfold band degeneracy. (c) The schematic diagram of a nodal line semimetal with a nodal line in the bulk and drumhead-like states on the surface.[14]

invariants within the material. This implies that they are robust against backscattering and can maintain strong electrical conductivity and catalytic activity even when faced with complex, strong acid/strong base electrolyte corrosion, or even the adsorption of small amounts of non-magnetic surface impurities and intrinsic defects. This fundamentally challenges the conventional wisdom that "defects easily become recombination centers and sources of deactivation" in traditional catalysts, significantly enhancing the catalyst's cycle life and robustness under real-world operating conditions.

To further validate the above theory, experimental and computational materials science has made significant progress in recent years regarding defects and surface engineering in various topological semimetals. Weyl semimetals are the first topological material system to be demonstrated as exhibiting a HER catalytic enhancement effect. Studies have shown that in light-driven HERs, the surface Fermi arcs of materials such as NbP and TaP serve as excellent electron transfer pathways. NbP exhibits a hydrogen production rate as high as 3,520 μMol/g, demonstrating exceptionally high intrinsic catalytic activity.[15]

Similar to Weyl semimetals, Dirac semimetals possess a double Fermi arc, offering more possibilities for catalysis. As shown in Figure 9.4, the figure compares the evolution of orbital interactions and Dirac point positions in pristine and defect-rich Nb_2S_2C before and after hydrogen adsorption. Before hydrogen atom adsorption, the Dirac point and associated TSSs of the system are located approximately –0.2 eV below the Fermi level; these states are primarily contributed by the highly chemically active S-3p_x and S-3p_y orbitals. Upon hydrogen adsorption onto the surface, these S-3p orbitals interact strongly with the hydrogen atom's orbitals, leading to significant charge accumulation on the H atom and the in-plane S-3p_x and S-3p_y orbitals, while the S-3p_z orbital perpendicular to the surface becomes charge-depleted. This spatial redistribution of charge greatly promotes the HER. Evolution of Dirac point

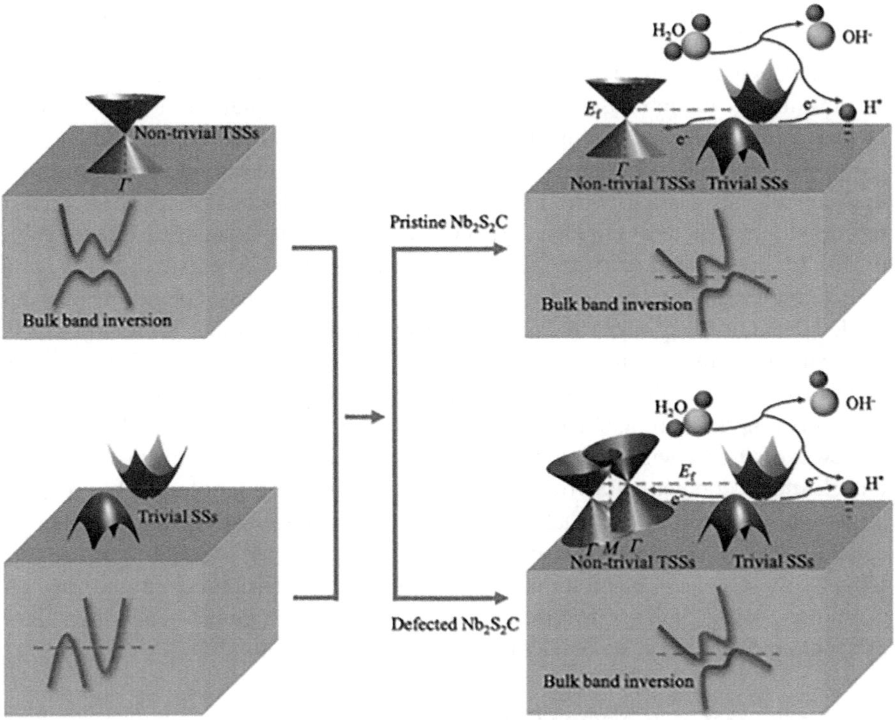

FIGURE 9.4 The synergistic effect between non-trivial topological and trivial surface states for activating the electrocatalytic HER activity in Dirac semimetals.[16]

positions following the introduction of sulfur vacancy defects (Pristine vs. Defect-rich) compared to the pristine Nb_2S_2C lattice, the introduction of S vacancy defects (Defect-rich) effectively shifts the position of the Dirac point. This band structure rearrangement, caused by the introduction of trivial defect states, further optimizes the ΔG_H^* of hydrogen adsorption, resulting in higher catalytic activity than that of the perfect lattice.[16]

As another example, nodal-line semimetals have demonstrated immense potential in the field of catalysis, surpassing that of traditional materials. The unique electronic structure of nodal-line semimetals in momentum space leads to highly distinctive 2D drumhead surface states. Compared to the 1D Fermi arcs in Weyl semimetals, these 2D surface states can generate[18] exceptionally large peaks in the surface electron DOSs near the Fermi level, providing extremely abundant and highly reactive pathways for molecular adsorption and charge transfer in electrochemical reactions. This physical advantage, stemming from the microscopic topological electronic structure, directly translates into remarkable macroscopic electrocatalytic performance. Taking the typical Dirac nodal-line semimetal $PtSn_4$ as an example, owing to the highly robust topological node-line features, $PtSn_4$ achieves a benchmark current density of 10 mA/cm^2 at an extremely low overpotential of just 37 mV. Its catalytic activity not only far exceeds that of conventional Pt foil but is even comparable to commercial

Pt/C catalysts. This strongly demonstrates that leveraging the intrinsic high activity of topological drumhead surface states can significantly reduce the amounts of precious metals used while achieving exceptional catalytic conversion efficiency.[44]

In summary, the macroscopic energy conversion performance of low-dimensional quantum materials is fundamentally dictated by their microscopic structural and electronic configurations. Overcoming the inherent physicochemical bottlenecks of traditional catalysts requires a shift from passive material selection to proactive, multidimensional structural engineering. As discussed, the layer number effect enables precise tuning of band structures and the maximization of exposed active sites, while defect engineering breaks the chemical inertness of perfect basal planes by introducing localized mid-gap states and optimizing adsorption energetics. More importantly, these structural modulations do not operate in isolation. Their synergistic interplay, coupled with strain, doping, and interfacial effects, provides a vast parameter space for catalyst optimization. When these classical geometric and chemical engineering strategies intersect with the emerging field of topological quantum materials, they unlock unprecedented catalytic paradigms. By harnessing topologically protected, high-mobility surface states that are remarkably robust against environmental degradation, researchers can transcend traditional catalytic scaling relationships. Ultimately, the deep integration of multidimensional structural engineering and quantum topology will serve as the cornerstone for designing next-generation, highly efficient, and durable sustainable energy conversion systems.

9.3 CATALYTIC MECHANISMS

9.3.1 Fundamental Kinetics in Catalytic Reactions

9.3.1.1 Adsorption and Desorption

Adsorption is the process by which reactant molecules achieve surface enrichment, orientation, and activation via physical interactions or chemical bonding at the active sites of a catalyst. Among these, chemisorption can directly cause the cleavage or weakening of intramolecular chemical bonds and serves as an essential prerequisite for the occurrence of surface reactions. Adsorption strength is typically quantified by the adsorption energy (E_{ad}), which is defined as $E_{ad} = E_{total} - \left(E_{substrate} + E_{adsorbate}\right)$. Here, E_{total}, $E_{substrate}$, and $E_{adsorbate}$ denote the total energy of the adsorption system, the energy of the substrate, and the energy of the isolated adsorbate molecule, respectively. A negative E_{ad} corresponds to exothermic adsorption; its absolute value reflects the adsorption strength and is closely correlated with catalytic activity. Desorption is the key step where reaction products desorb from the catalyst surface and return to the bulk phase. Reactant molecules first adsorb onto the catalyst surface, undergo activation to initiate chemical reactions, and finally, the products desorb from the surface, thus completing a catalytic cycle.

Topological insulators (TIs) such as Bi_2Se_3 and Bi_2Te_3 exhibit unique surface states whose spin-momentum locking prevents backscattering, creating unusually stable and active adsorption sites.[39] The Dirac-cone electronic structure of TI surfaces allows frontier electrons to interact with adsorbates at lower energy costs than conventional metals. DFT studies show that the adsorption energy of O* and OH* intermediates on Bi_2Se_3 (111) surfaces lies within the Sabatier optimum window for ORR.[20] Furthermore, the topologically protected surface states remain robust against moderate structural perturbations, ensuring consistent adsorption behavior across different surface morphologies. Adsorption and desorption in catalytic processes are the primary steps of heterogeneous catalysis.

9.3.1.2 Stability of Reaction Intermediates

The 1D topological edge states of 2D TIs (e.g., WTe_2 monolayers) provide a quasi-continuous density of electronic states that can selectively stabilize polar reaction intermediates through long-range electrostatic interactions.[19] Unlike conventional surface defects, topological edge states are non-local and immune to single-atom disorder, resulting in more reproducible intermediate stabilization across different synthesis batches.[20] Catalytic selectivity is predominantly governed by the stability of reaction intermediates. In intricate multi-step reaction systems, the reaction pathway is directed by the relative stability of distinct reaction intermediates. In the CO_2 reduction reaction, the branching of the reaction pathway begins with the key intermediates formed after CO_2 activation. Following the activation of CO_2 molecules on the catalyst surface, two critical intermediates can be formed via proton-coupled electron transfer: carbon-coordinated COOH and oxygen-coordinated OCHO. The relative stability of these two intermediates dictates the direction of the subsequent reaction. That is, stabilization of OCHO favors the production of formic acid (HCOOH), whereas stabilization of COOH leads to the formation of CO or further reduced products.[21]

9.3.1.3 Lowering of Activation Energy

The Dirac-fermion nature of charge carriers in TSSs enables near-barrierless electron transfer, effectively lowering activation energy for redox-driven catalytic steps. In photocatalytic applications, the linear dispersion of topological bands facilitates hot-carrier generation and rapid charge migration to reaction sites before recombination occurs. The connection between topological protection and kinetic acceleration bridges condensed-matter physics and heterogeneous catalysis in a fundamentally new way.[24]

The lowering of activation energy is the kinetic basis of catalysis. According to the Arrhenius equation $k = Ae^{-Ea/RT}$, the reaction rate constant is exponentially related to the activation energy. An excellent catalyst can significantly lower the activation energy by providing an alternative reaction pathway. Quantitative expression of the lowering of activation energy: the ratio of the catalyzed reaction rate k_{cat} to the uncatalyzed reaction rate k_{uncat} satisfies. When the activation energy of a catalyzed reaction is 50 kJ/mol lower than that of the uncatalyzed reaction, the reaction rate at room temperature can increase by approximately five orders of magnitude. In 2022, Ma et al. reported that constructing a 0D/2D heterostructure from 0D graphene quantum dots (GQDs) and 2D MXene ($Ti_3C_2O_2$) significantly reduces the activation energy barrier for OER, demonstrating an innovative strategy for lowering activation energy through the synergistic effect of low-dimensional quantum materials.[23]

9.3.1.4 Sabatier Principle

The Sabatier principle was proposed in the early 20th century by Nobel laureate Paul Sabatier. Its core concept is that there exists an optimal range of interaction strength between a catalyst and the reactants (or reaction intermediates). If the interaction is too weak, the reactants cannot be effectively adsorbed on the catalyst surface for further activation, limiting the catalytic reaction rate. If the interaction is too strong, the products are difficult to desorb from the surface, and the catalyst is prone to being "poisoned" by intermediates or having its surface occupied, again restricting the reaction rate. Only when the interaction strength is in an appropriate range can the reactants be smoothly adsorbed and activated, and the products desorb in a timely manner, allowing the reaction rate to reach its maximum. This results in the classic volcano-shaped curve in the relationship between catalytic activity and adsorption energy. For ORR, the adsorption energy of the OH intermediate serves as the descriptor. Pt is located near the volcano peak, whereas Au binds too weakly (resulting in a large overpotential) and Ni binds too strongly (leading to oxygen poisoning).

9.3.1.5 D-Band Center or Electronic State Distribution Regulation

The d-band center theory is an important theoretical framework for understanding the catalytic activity of transition metals. Nørskov and colleagues proposed that the d-band center energy (ε_d) of a transition metal surface determines its adsorption strength for reaction intermediates. The higher the d-band center, the less the antibonding states are filled, and the stronger the adsorption. This theory provides critical theoretical guidance for the design and optimization of catalysts.

In transition-metal-based topological semimetals (e.g., $Co_3Sn_2S_2$ magnetic Weyl semimetal), the interplay between narrow d-bands and topological Weyl nodes creates a hybridized electronic structure that modifies the d-band center position in ways inaccessible to trivial materials.[26] Tuning SOC strength through elemental substitution shifts the Weyl node energy and simultaneously adjusts the d-band center, providing an additional degree of freedom beyond conventional d-band engineering.[27] X-ray absorption spectroscopy has revealed orbital-specific electronic structures of $Co_3Sn_2S_2$, while topological surface states have been proposed to contribute to enhanced catalytic activity.[26]

9.3.1.6 Charge Transfer and Local Electronic Structure

The essence of catalytic reactions lies in electron transfer, which drives the breaking and formation of chemical bonds. The efficiency and direction of this process fundamentally depend on how well the electronic structures of the catalyst and the reactant molecules match.[29] A deep understanding of electron behavior is therefore critical for optimizing catalytic efficiency. Low-dimensional quantum materials, with their unique quantum confinement effects, serve as ideal platforms for studying charge transfer and local electronic structure. When the material size is reduced to the nanometer scale, the coordination environment of surface atoms becomes unsaturated, triggering a reconstruction of the local electronic structure. This is manifested as a redistribution of the electronic DOSs and the formation of defect levels, thereby significantly enhancing the material's charge transfer capability.[30]

This feature is particularly prominent in photocatalytic systems based on low-dimensional quantum materials. Taking perovskite quantum dots as an example, there is a dynamic coupling between the evolution of photogenerated excitons and the behavior of surface reactant molecules. Studies have shown that molecules adsorbed on the quantum dot surface can form "quasi-ligand" coordination and undergo "crosstalk" with photogenerated excitons. This exciton–molecule interaction can simultaneously suppress nonradiative relaxation pathways such as radiative recombination and Auger recombination, and reduce the exciton binding energy of the system. As a result, the reaction activation energy is effectively lowered, and the catalytic efficiency is significantly improved. Further research has revealed that surface modification of quantum dots through ligand engineering can regulate exciton dissociation and charge transfer kinetics.[31] For example, modifying $CsPbBr_3$ quantum dots with ferrocenecarboxylic acid enhances exciton dissociation and charge transfer through its dielectric shielding effect and electron transport capability, ultimately boosting photocatalytic performance. Therefore, precise control over the local electronic structure of low-dimensional quantum materials and the mechanisms of interfacial charge transfer is a key scientific approach for developing highly efficient catalytic systems.[32]

A hallmark of topological materials is the absence of backscattering for surface/edge electrons due to time-reversal symmetry protection. In catalytic contexts, this translates to ultralow charge transfer resistance across topological material–electrolyte interfaces. The suppression of backscattering also prevents charge trapping at grain boundaries, a common problem in polycrystalline photocatalysts. Consequently, topological electrocatalysts maintain high turnover frequencies even in large-area electrode configurations. In 1T′-$MoTe_2$ single crystals, EIS measurements reveal a contact resistance as low as 0.3 Ω, which corresponds to ultrafast reaction kinetics.

9.3.2 Active Sites in Low-Dimensional Materials

9.3.2.1 Basal Plane Sites

Basal plane sites are the most abundant intrinsic sites in 2D materials. Taking TMDs (e.g., MoS_2) as an example, the basal plane consists of a hexagonally close-packed arrangement of metal atoms and chalcogen atoms, and the perfect lattice is typically catalytically inert. However, strategies such as defect engineering, phase transformation, or external strain can effectively activate the basal plane atoms, endowing them with catalytic reactivity.

In 2H-MoS_2 and related layered compounds, the catalytically inert basal plane becomes activated upon transition to a topological phase (e.g., 1T′ phase). The metallic 1T′ phase of MoS_2 supports topological edge states and simultaneously lowers the HER overpotential across the entire basal plane from >500 mV to <100 mV.[33] This phase-topology coupling was unambiguously demonstrated by in situ Raman and STM studies showing that the HER activity scales with topological domain area, not merely edge length. This implies that topological phase engineering can transform previously wasted basal plane area into high-density active site arrays.[34]

9.3.2.2 Edge Sites

Edge sites have long been considered the primary active sites in 2D materials. Taking the HER on TMDs (such as MoS_2) as an example, both theoretical and experimental studies have confirmed that the catalytic activity mainly originates from the sulfur atoms at the edges, while the large-area basal plane atoms exhibit inertness. Hinnemann et al., through DFT calculations combined with electrochemical experiments, were the first to confirm that the HER activity of low-dimensional MoS_2 materials originates solely from the edge sites. The unsaturated coordination of Mo and S atoms at the edges achieves an optimal hydrogen adsorption free energy comparable to that of platinum, whereas the basal plane sites of MoS_2, with their coordinatively saturated atoms, are intrinsically catalytically inert and cannot effectively adsorb or activate hydrogen intermediates.[45] Jaramillo et al. experimentally synthesized MoS_2 nanostructures with controllable edge-site densities, providing the first quantitative experimental demonstration that catalytic activity is linearly correlated with the number of edge sites, thereby validating the theoretical predictions.[34] While conventional edge sites in 2D catalysts (e.g., MoS_2) degrade under reaction conditions due to oxidation or dissolution, topological edge states of quantum spin Hall insulators (e.g., WTe_2 monolayer) are protected by bulk topology and cannot be destroyed by non-magnetic perturbations. The edge states of WTe_2 are not affected by the specific edge configuration and can exist stably. This inherent durability makes topological edge states particularly valuable for long-term electrocatalytic applications requiring thousands of operating hours.[19,20]

9.3.2.3 Vacancy Defects

Vacancy defects are the most common type of point defects in low-dimensional quantum materials, including sulfur vacancies, oxygen vacancies, and so on. The introduction of vacancies creates defect levels within the bandgap, altering the electronic structure and carrier dynamics of the material. In catalytic reactions, vacancies can serve as preferential adsorption sites for reactant molecules and act as bridges for charge transfer, thereby facilitating the catalytic reaction.

Specifically, vacancy defects in 2D materials can drive topological phase transitions that unlock entirely new catalytic pathways. Lopes et al. recently demonstrated a general, material-independent mechanism by which atomic vacancies induce topological phase transitions in 2D semiconductors that are otherwise topologically trivial.[46] In MXene materials (e.g., $Ti_3C_2T_x$), selective surface-group removal creates local strain fields that may invert the bandgap and induce a topological phase, as suggested by first-principles calculations. The resulting topological domain boundaries concentrate charge density and exhibit substantially lower CO_2-to-CO conversion barriers than defect-free regions. For MXenes (e.g., $V_{1.33}C(OH)_2$), DFT studies reveal that ordered metal atomic vacancies produce strong O-H bonds that prevent H detachment, with predicted limiting potentials as low as –0.32 to –0.15 V.[35] Crucially, these vacancy-induced topological domains may self-organize into percolating networks during synthesis, potentially providing macroscopic catalytic enhancement without precise atomic-scale engineering.

9.3.2.4 Heteroatom Doping Site and Single-Atom Anchoring Sites

Heteroatom doping provides a powerful lever to position the Fermi level optimally within the TSS band for target reactions. By introducing heteroatoms such as N, B, P, S, or transition metals, the local electronic structure, bandgap, and carrier concentration of the material can be effectively modified, thereby optimizing catalytic performance.[39] In carbon-based quantum dots, this strategy has been shown to significantly enhance electrocatalytic activity. Studies indicate that in nitrogen-doped GQDs, the pyridinic nitrogen configuration induces charge redistribution and a higher spin density on adjacent carbon atoms, making them active centers for the ORR.[47]

The recently developed single-atom catalysis represents atomic-level precision in active site design, anchoring isolated metal atoms onto a support surface to achieve nearly 100% atom utilization. Low-dimensional quantum materials, with their high specific surface area and abundant defect sites, serve as ideal supports. A typical example is the anchoring of Fe single atoms on N-doped graphene, yielding a catalyst with a half-wave potential of 0.911 V for the ORR, outperforming commercial Pt/C.[48] Furthermore, atomically precise metal clusters exhibit molecule-like electronic structures due to quantum confinement effects, leading to unique catalytic behavior. Doping engineering and ligand design can further optimize their catalytic performance, pushing catalyst design toward precise atomic-scale regulation.

TI surfaces provide unique platforms for immobilizing single-atom catalysts (SACs) owing to their topologically protected surface states and intrinsic charge inhomogeneity near the Dirac point. DFT calculations demonstrate that single Pt atoms adsorbed on the Bi_2Se_3 (111) surface exhibit significantly higher diffusion barriers compared with those on conventional carbon supports,[49] which originates from strong orbital hybridization between Pt 5d states and TSSs of Bi_2Se_3. Such strong metal–support interactions enable ultrahigh SAC loadings without aggregation, far exceeding the loading limit typically achieved on traditional substrates. Moreover, the topological substrate facilitates efficient charge transfer from the bulk to the anchored single Pt sites, which substantially enhances the electrocatalytic activity toward methanol oxidation.

9.4 CHALLENGES AND DEVELOPMENT DIRECTIONS

9.4.1 Challenges

The practical deployment of low-dimensional quantum materials in catalysis is fundamentally constrained by their limited stability under realistic operating conditions. The intrinsic advantages of these systems, such as high surface-to-volume ratio, quantum confinement, and abundant unsaturated sites, also render them highly susceptible to external perturbations. Under coupled stimuli, including illumination, applied bias, and reactive electrolytes, degradation processes such as ion migration, surface oxidation, phase transformation, and interfacial delamination are significantly accelerated. For example, in quantum-confined systems, reduced coordination and weakened lattice stability amplify structural reconstruction and chemical corrosion, while strong surface sensitivity leads to rapid deterioration of electronic states and catalytic functionality. In addition, dynamic processes such as dissolution–redeposition equilibrium and support corrosion further complicate long-term stability, particularly in harsh electrochemical environments, where the delicate balance between structural integrity and surface reactivity becomes increasingly difficult to maintain.

Beyond stability, a deeper challenge lies in achieving precise control and fundamental understanding of active sites in low-dimensional quantum materials. Unlike conventional catalysts, their catalytic behavior is governed not only by local atomic configurations but also by quantum effects such as band structure renormalization, edge states, and many-body interactions. However, defect engineering in these systems remains intrinsically stochastic, making it difficult to deterministically regulate defect type, density, and spatial distribution at scale. More critically, catalytic active sites are dynamically reconstructed under reaction conditions, with their electronic structure and local coordination continuously evolving, often accompanied by quantum state transitions or charge redistribution. This complexity is further compounded by the discrepancy between model systems and practical catalysts: insights derived from idealized low-dimensional structures (e.g., single-crystal surfaces or isolated nanosheets) often fail to capture the multiscale heterogeneity, high coverage effects, and environment-induced electronic reconstruction present in real catalytic systems.

9.4.2 Future Perspectives

Future progress will rely on advancing from nanoscale engineering toward atomically and quantum-state-resolved design of catalytic materials. In low-dimensional quantum systems, catalytic activity is intrinsically linked to the precise electronic structure, including d-band centers, Dirac states, or topologically protected surface states. Thus, the next frontier lies in achieving deterministic control over atomic arrangement and quantum states simultaneously. Strategies such as SACs, heterostructure engineering, and interfacial electronic coupling offer promising routes to tailor charge distribution and optimize adsorption energetics. However, key challenges remain in stabilizing atomically dispersed active sites at high loading while preserving quantum coherence and preventing aggregation or phase degradation. Meanwhile, artificial intelligence is poised to play a transformative role by uncovering hidden correlations between quantum electronic descriptors and catalytic performance, enabling accelerated discovery of low-dimensional quantum catalysts beyond conventional chemical intuition.

Equally critical is the development of integrated, multiscale research frameworks that combine high-throughput computation, automated experimentation, and operando characterization. For low-dimensional quantum materials, capturing the dynamic evolution of quantum states under working conditions is essential. Advanced operando techniques, particularly when coupled with machine learning, can track real-time changes in electronic structure, coordination environment, and phase behavior, thereby establishing direct structure–property relationships. When integrated with high-throughput screening and active learning strategies, these approaches enable a closed-loop paradigm for catalyst design, bridging the gap between theoretical prediction and practical implementation. Ultimately, such convergence is expected to unlock the full potential of low-dimensional quantum materials, enabling the rational design of catalytic systems that are not only highly active but also robust and adaptive under realistic operating conditions.

REFERENCES

[1] A. Fujishima and K. Honda, Electrochemical photolysis of water at a semiconductor electrode. *Nature* 238, 37–38 (1972).

[2] P. De Luna, C. Hahn, D. Higgins, S. A. Jaffer, T. F. Jaramillo and E. H. Sargent, What would it take for renewably powered electrosynthesis to displace petrochemical processes? *Science* 364, eaav3506 (2019).

[3] J. M. Doña-Rodríguez and E. Pulido Melián, Nano-photocatalytic materials: Possibilities and challenges. *Nanomaterials* 11, 688 (2021).

[4] J. Di, J. Xia, H. Li and Z. Liu, Freestanding atomically-thin two-dimensional materials beyond graphene meeting photocatalysis: Opportunities and challenges. *Nano Energy* 35, 79–91 (2017).

[5] A. P. Alivisatos, Semiconductor clusters, nanocrystals, and quantum dots. *Science* 271, 933–937 (1996).

[6] M. Chhowalla, H. S. Shin, G. Eda, L.-J. Li, K. P. Loh and H. Zhang, The chemistry of two-dimensional layered transition metal dichalcogenide nanosheets. *Nature Chemistry* 5, 263–275 (2013).

[7] Y. Ying, K. Fan, Z. Lin and H. Huang, Facing the "cutting edge:" Edge site engineering on 2D materials for electrocatalysis and photocatalysis. *Advanced Materials* 37, 2418757 (2025).

[8] H. Wang, W. Liu, X. He, P. Zhang, X. Zhang and Y. Xie, Toward an excitonic perspective on low-dimensional semiconductors for photocatalysis. *Journal of the American Chemical Society* 142, 14007–14022 (2020).

[9] H. Wang, L. Zhang, Z. Chen, J. Hu, S. Li, Z. Wang, J. Liu and X. Wang, Semiconductor heterojunction photocatalysts: design, construction, and photocatalytic performances. *Chemical Society Reviews* 43, 5234–5244 (2014).

[10] A. Kudo and Y. Miseki, Heterogeneous photocatalyst materials for water splitting. *Chemical Society Reviews* 38, 253–278 (2008).

[11] T. Takata, J. Jiang, Y. Sakata, M. Nakabayashi, N. Shibata, V. Nandal, K. Seki, T. Hisatomi and K. Domen, Photocatalytic water splitting with a quantum efficiency of almost unity. *Nature* 581, 411–414 (2020).

[12] J. Yang, Y. Wu, Z. Yu, A. Politano, D. Bukhvalov, A. Cupolillo, H. Feng, X. Zhang and Y.-W. Zhang, Exploring topological materials for hydrogen evolution reaction: insights from density functional theory. *Energy Materials* 6, 600029 (2026).

[13] J. Xie, H. Zhang, S. Li, R. Wang, X. Sun, M. Zhou, J. Zhou, X. W. Lou and Y. Xie, Defect-rich MoS_2 ultrathin nanosheets with additional active edge sites for enhanced electrocatalytic hydrogen evolution. *Advanced Materials* 40, 5807–5813 (2013).

[14] L. Wang, Y. Yang, J. Wang, W. Liu, Y. Liu, J. Gong, G. Liu, X. Wang, Z. Cheng and X. Zhang, Excellent catalytic performance toward the hydrogen evolution reaction in topological semimetals. *EcoMat* 5, e12316 (2023).

[15] C. R. Rajamathi, U. Gupta, N. Kumar, H. Yang, Y. Sun, V. Sü, C. Shekhar, M. Schmidt, H. Blumtritt, P. Werner, B. Yan, S. Parkin, C. Felser and C. N. R. Rao, Weyl semimetals as hydrogen evolution catalysts. *Advanced Materials* 29, 1606202 (2017).

[16] Q. Yang, C. Le, G. Li, T. Heine, C. Felser and Y. Sun, Enhancement of basal plane electrocatalytic hydrogen evolution activity via joint utilization of trivial and non-trivial surface states. *Applied Materials Today* 22, 100921 (2021).

[17] K. Zollner and J. Fabian, Heterostructures of graphene and topological insulators Bi_2Se_3, Bi_2Te_3, and Sb_2Te_3. *Physica Status Solidi (B) - Basic Solid State Physics* 258, 2000081 (2021).

[18] C. Fan, K. Sakamoto and P. Krüger, Carrier doping of Bi_2Se_3 surface by chemical adsorption—A DFT study. *Applied Surface Science* 643, 158699 (2024).

[19] L. Peng, Y. Yuan, G. Li, X. Yang, J.-J. Xian, C.-J. Yi, Y.-G. Shi and Y.-S. Fu, Observation of topological states residing at step edges of WTe_2. *Nature Communications* 8, 659 (2017).

[20] Y. Maximenko, Y. Chang, G. Chen, M. R. Hirsbrunner, W. Swiech, T. L. Hughes, L. K. Wagner and V. Madhavan, Nanoscale studies of electric field effects on monolayer $1T'$-WTe_2. *NPJ Quantum Materials* 7, 29 (2022).

[21] M. Zhang, H. Li, X. Lin, Z. Zhang, H. Xue, Y. Zhang, Y. Deng and Y. Jiang, Steering carbon dioxide electroreduction toward formic acid by Sn-induced Cu0 stabilization and optimized *OCHO intermediate adsorption. *Journal of Colloid and Interface Science* 716, 140356 (2026).

[22] Y. Zhu, Q. Han and Y. Zhou, Synergistic selenium vacancies and bismuth metal centers on Bi2Se3 for enhanced photocatalytic CO_2 reduction. *Nanoscale Advances* 7, 6640–6645 (2025).

[23] Y. Ma, Y. An, Z. Xu, L. Cheng and W. Yuan, Activating lattice oxygen of two-dimensional MnXn–$1O_2$ MXenes via zero-dimensional graphene quantum dots for water oxidation. *Science China Materials* 65, 3053–3061 (2022).

[24] W. Liu, X. Zhang, W. Meng, Y. Liu, X. Dai and G. Liu, Theoretical realization of hybrid Weyl state and associated high catalytic performance for hydrogen evolution in NiSi. *iScience* 25 (2022).

[25] T. W. Lau, Q. Lei and J. Yin, Facet engineering of Weyl semimetals for efficient hydrogen evolution reaction. *Materials Today Catalysis* 8, 100091 (2025).

[26] J. Dong, S. Chen, X. Wang, Y. Mao, X. Jing, Z. Ouyang, J. Ding, J. Tian, Y. Ji, Y. Li, W.-J. Shi, Y. Han and W. Zhang, Tuning the band structure of $Co_3Sn_2S_2$ topological semimetal for oxygen evolution in alkaline electrolytes. *ACS Applied Materials & Interfaces* 17, 64392–64402 (2025).

[27] B. Hammer and J. K. Nørskov, Electronic factors determining the reactivity of metal surfaces. *Surface Science* 343(1–3), 211–220 (1995).

[28] T. Qi, Y. Shao, C. Zhang, D. Wang, F. Han, Q. Jin, Z. Liu, X. Zhang, L. Li and L. Wu, Synergistic catalytic mechanism of constituent metal atoms in high-entropy alloy catalysts for Li – S batteries. *Chemical Engineering Journal* 31, 175721 (2026).

[29] C. Du, J. Sheng, F. Zhong, Y. He, H. Liu, Y. Sun and F. Dong, Boosting exciton dissociation and charge transfer in $CsPbBr_3$ QDs via ferrocene derivative ligation for CO_2 photoreduction. *Proceedings of the National Academy of Sciences of the United States of America* 121, e2315956121 (2024).

[30] X.-L. Fan, Y. Yang, P. Xiao and W.-M. Lau, Site-specific catalytic activity in exfoliated MoS2 single-layer polytypes for hydrogen evolution: basal plane and edges. *Journal of Materials Chemistry A* 2, 20545–20551 (2014).

[31] M. Yan, Y. Jin, X. Hou, Y. Guo, A. Tsaturyan, A. Makarova, … E. Voloshina, Topological quasi-2D semimetal Co_3Sn2S_2: Insights into electronic structure from NEXAFS and resonant photoelectron spectroscopy. *The Journal of Physical Chemistry Letters*, 12(40), 9807–9811 (2021).

[32] C. Du, J. Sheng, F. Zhong, Y. He, H. Liu, Y. Sun, and F. Dong, Boosting exciton dissociation and charge transfer in $CsPbBr_3$ QDs via ferrocene derivative ligation for CO2 photoreduction. *Proceedings of the National Academy of Sciences*, 121(9), e2315956121 (2024).

[33] J. F. Vliem, S. A. Cayan, R. Reho, A. R. Botello-Méndez, P. Geiregat, Z. Zanolli and D. Vanmaekelbergh, Role of surface bands in the photogeneration, cooling, and recombination of charge carriers in two-dimensional Bi_2Se_3. *ACS Nano* 19, 17261–17272 (2025).
[34] T. F. Jaramillo, K. P. Jørgensen, J. Bonde, J. H. Nielsen, S. Horch and I. Chorkendorff, Identification of active edge sites for electrochemical H_2 evolution from MoS_2 nanocatalysts. *Science* 317, 100–102 (2007).
[35] H. Xue, Y. Chen, L. Zhu and A. Du, Enhanced catalytic activity of i-MXenes for CO_2 reduction reaction by ordered metal atomic vacancies: A DFT study. *Surfaces and Interfaces* 56, 105535 (2025).
[36] C. Tsai, H. Li, S. Park, J. Park, H. S. Han, J. K. Nørskov, X. Zheng and F. Abild-Pedersen, Electrochemical generation of sulfur vacancies in the basal plane of MoS_2 for hydrogen evolution. *Nature Communications* 8, 15113 (2017).
[37] C. T. Campbell and C. H. Peden, Oxygen vacancies and catalysis on ceria surfaces. *Science* 309, 713–714 (2005).
[38] H. Li, J. Li, Z. Ai, F. Jia and L. Zhang, Oxygen vacancy-mediated photocatalysis of BiOCl: reactivity, selectivity, and perspectives. *Angewandte Chemie International Edition* 57, 122–138 (2018).
[39] H. Zhang, C.-X. Liu, X.-L. Qi, X. Dai, Z. Fang and S.-C. Zhang, Topological insulators in Bi_2Se_3, Bi_2Te_3 and Sb_2Te_3 with a single Dirac cone on the surface. *Nature Physics* 5, 438–442 (2009).
[40] C.-C. Wang, A.-Y. Lo, M.-C. Cheng, Y.-S. Chang, H.-C. Shih, F.-S. Shieu, T.-H. Tseng and H.-T. Tsai, Enhanced electrical properties of amorphous In-Sn-Zn oxides through heterostructuring with Bi_2Se_3 topological insulators. *Scientific Reports* 14, 195 (2024).
[41] S. Jayachitra, P. Ravi, P. Murugan and M. Sathish, Supercritically exfoliated Bi_2Se_3 nanosheets for enhanced photocatalytic hydrogen production by topological surface states over TiO_2. *Journal of Colloid and Interface Science* 605, 871–880 (2022).
[42] K. F. Mak, C. Lee, J. Hone, J. Shan, and T. F. Heinz, Atomically thin MoS_2: A new direct-gap semiconductor. *Physical Review Letters*, 105, 136805 (2010).
[43] J. Xie, H. Zhang, S. Li, R. Wang, X. Sun, M. Zhou, J. Zhou, X. W. D. Lou, and Y. Xie, Defect-rich MoS_2 ultrathin nanosheets with additional active edge sites for enhanced electrocatalytic hydrogen evolution. *Advanced Materials*, 25, 5807–5813 (2013).
[44] L. Wang, Y. Yang, J. Wang, W. Liu, Y. Liu, J. Gong, G. Liu, X. Wang, Z. Cheng, and X. Zhang, Excellent catalytic performance toward the hydrogen evolution reaction in topological semimetals. *EcoMat*, 5, e12316 (2023).
[45] B. Hinnemann, P. G. Moses, J. Bonde, K. P. Jørgensen, J. H. Nielsen, S. Horch, I. Chorkendorff, and J. K. Nørskov, Biomimetic hydrogen evolution: MoS_2 nanoparticles as catalyst for hydrogen evolution. *Journal of the American Chemical Society*, 127, 5308–5309 (2005).
[46] E. V. C. Lopes, F. C. de Lima, C. Lewenkopf, and A. Fazzio, Engineering quantum phases in two dimensions via vacancy-induced electronic reconstruction. *arXiv preprint arXiv:2603.17122* (2026).
[47] J. Yuan, X. B. Yang, H. F. Wang, Y. Cao, H. Wang, G. Yang, and H. Yu, Pyridinic-N dominates ORR performance: decoding configuration-sensitive electrocatalysis of N-doped graphene quantum dots. *ACS Catalysis*, 15, 18112–18122 (2025).
[48] J. Wang, J. Liu, Y. Qiu, Y. Wang, and X. Cui, Bifunctional oxygen electrocatalytic performance of atomically dispersed Fe anchored on N-doped graphene. *Journal of Inorganic Materials*, 41, 814–822 (2026).
[49] J. Xiao, L. Kou, C. Y. Yam, T. Frauenheim, and B. Yan, Toward rational design of catalysts supported on a topological insulator substrate. *ACS Catalysis*, 5, 7063–7067 (2015).

Index

Pages in *italics* refer to figures.

For Product Safety Concerns and Information please contact our EU representative GPSR@taylorandfrancis.com Taylor & Francis Verlag GmbH, Kaufingerstraße 24, 80331 München, Germany

Batch number: 10837190

Printed by Printforce, the Netherlands